Václav Čížek

DISCRETE FOURIER TRANSFORMS AND THEIR APPLICATIONS

ADAM HILGER Ltd
BRISTOL and BOSTON

British Library Cataloguing in Publication Data

Čížek, Václav
Discrete Fourier transforms and their applications.
1. Fourier transformations
I. Title II. Diskretní Fourierova transformace a její použití. *English*
515.7′23 QA403.5

ISBN 0-85274-800-0

Translation: Jaroslav Jirát

Adam Hilger Ltd
Techno House, Redcliffe Way, Bristol BS1 6NX, England
PO Box 230, Accord, MA 02018, USA
Published in co-edition with SNTL, Publishers of Technical Literature, Prague

Printed in Czechoslovakia

CONTENTS

PREFACE

The steady growth in the use of Fourier transforms by workers in many fields, primarily in the natural and technical sciences, makes it necessary to master suitable methods for their practical computation using computers. This book has arisen from the need to provide users of Fourier transforms with a review of basic knowledge on the discrete Fourier transform (DFT) and its relation to the Fourier transform and Fourier series. This knowledge is a prerequisite for the most efficient use of DFTs and the proper interpretation of results so obtained.

The material to be found in this book was used in several graduate courses and lectures for practising workers. It provides basic knowledge for the practical use of the DFT and enables further study of the literature on its properties and applications.

The book has been published, due to the initiative of Adam Hilger Ltd, as a coedition with the Czech publishing house SNTL. It represents an expansion and updating of the book *Discrete Fourier Transforms and their Applications,* published by SNTL in 1981 in the Czech language.

Finally, the author would like to thank the management of the Institute of Radioengineering and Electronics of the Czechoslovak Academy of Sciences for their understanding in the writing of this book, and colleagues of the Department of the Theory of Electronic Systems for their encouragement in preparing the manuscript.

Prague, January 1985 V. Čížek

1. INTRODUCTION

1.1 Origin and meaning of discrete Fourier transforms

Discrete Fourier transforms, better known to the general public as fast Fourier transforms, represent one of the computational methods which have undergone a rapid development since their appearance in 1965 – 1967. Discrete Fourier transforms as a means of computing Fourier transforms arose essentially from the needs of engineers and physicists, but primarily of electronics engineers. They developed further in the electronics field, and subsequently penetrated successfully into a number of other fields of science. This was due largely to an admirably fast algorithm for their computation published in 1965 by J. W. Cooley and J. W. Tukey, to a widespread need for computing Fourier transforms, and to developments in computers.

The term fast Fourier transform (FFT) was first used in 1967 [21] and its basic properties were first described in 1969 [24]. Of course, this does not mean that the relations which define the discrete Fourier transform were not used before. Relations which occur on writing out the discrete Fourier transform in its real and imaginary parts have been known since 1815, and these are generally called Bessel relations [68]. For a long time, these relations provided one of the most frequently used methods for numerical harmonic analysis: the Runge method.

1.2 Applications of discrete Fourier transforms

At present the discrete Fourier transform is one of the most advanced methods of discrete signal processing. Discrete Fourier transforms and the $\mathscr{Z}$ transforms constitute two basic techniques in that field. For numerical mathematics the discrete Fourier transforms and algorithms of their computation have made a valuable contribution and computer programs for their use are common.

Today one can hardly imagine computing Fourier transforms without the algorithms of the fast Fourier transform. One of the proofs of the exceptional efficiency of the technique is the fact that several manufacturers supply special dedicated computers (processors) for its computation or produce measuring instruments with these processors which permit the user to carry out spectral analysis in real time up to a frequency of 50 kHz. The discrete Fourier transform enables the user to create powerful methods for Fourier infrared spectroscopy and for nuclear spectroscopy. The microprocessor boom renders further such applications possible.

To illustrate the applications of the discrete Fourier transform we present, without claiming to be complete, a list of fields where it is used: communications and electronics (linear and nonlinear circuits and systems, signal processing, antennas, digital filters), acoustics and speech processing, radar and sonar, geophysics and seismology, machine dynamics, radioastronomy, oceanology, thermodynamics, analysis of control systems, biomedicine, hydrodynamics, spectrometry, optics and image processing, quantum physics, musicology.

1.3 Aim of the book

This book is intended to bring to the reader some basic information on discrete Fourier transforms. Limitations of size prevent a full review of properties and applications. Basic properties are covered, therefore, only from the point of view of their application to Fourier transforms. This topic is dealt with in Chapters 4 and 5.

As the major use of discrete Fourier transforms is for giving an approximate expression of Fourier transforms, Chapter 3 (on numerical computation of Fourier transforms) is included. Chapter 6 is devoted to basic principles of algorithms for computing discrete Fourier transforms. A classical application of discrete Fourier transforms to the calculation of a convolution is described in Chapter 7. Chapter 8 deals with a nonconventional application of the discrete Fourier transform to an approximate computation of the Hilbert transform.

The most frequent application of discrete Fourier transforms – spectral analysis – is completely omitted. This is a highly developed subject with specific methods which deserve separate treatment.

The book is written particularly for users of discrete Fourier transforms in the fields of engineering and some areas of natural sciences. For that reason, the presentation is chosen to be intelligible to workers in subjects other than communications, even though a signal–spectral interpretation might be more attractive to users in this discipline. The book, may, therefore, be useful even to graduate students in certain courses.

2. FOURIER SERIES

Discrete Fourier transforms are closely related to Fourier series and Fourier transforms. In order to demonstrate some of the links, we present in this chapter a concise review of some basic properties of Fourier series. As the delta distribution is often used in applications of the discrete Fourier transform, a review of its basic properties is also included. For the same reason, some properties of the Fourier and Hilbert transforms are presented at the end of the chapter.

The aim of this chapter is not a full discussion of properties, but just the establishment of a certain corpus of facts which should be borne in mind by any user of discrete Fourier transforms when studying this book, and when carrying out particular applications.

Such a collection of a large number of facts cannot, of course, contain proofs, and it is not even always possible to maintain the usual mathematical exactness. For a thorough discussion and a complete treatment, the reader must, therefore, consult some of the books cited in the text.

2.1 Basic properties of Fourier series

2.1.1 Definition of Fourier series

The Fourier series for a given function x (generally a complex one), periodic with a period P, $P > 0$, is defined as an infinite series

$$\text{(2-1)} \qquad \xi(t) = \sum_{k=-\infty}^{\infty} C_k \, e^{jk2\pi t/P}, \qquad -\infty < t < \infty,$$

with coefficients satisfying

$$\text{(2-2)} \qquad C_k = \frac{1}{P} \int_0^P x(t) \, e^{-jk2\pi t/P} \, dt, \qquad k = 0, \pm 1, \pm 2, \ldots$$

where the letter j indicates the imaginary unit $j = \sqrt{-1}$. The Fourier series is an expansion in terms of the orthogonal functions $e^{jk2\pi t/P}$, $k = 0, \pm 1, \pm 2, \ldots$, for which

$$\text{(2-3)} \qquad \int_0^P e^{jk2\pi t/P} \, e^{ji2\pi t/P} \, dt = \begin{cases} 0 & i \neq -k, \\ P, & i = -k. \end{cases}$$

As the coefficients C_k are determined just by values of the function x in a periodicity interval $[0, P)$ we can also construct a Fourier series for a function which is not periodic, but which is either defined only on this interval $[0, P)$, or which differs from zero only on this interval, i.e. for which $x_0(t) = 0, t > P, t < 0$. From such a function x_0, a periodic function x may be constructed using the rule

$$(2\text{-}4) \qquad x(t) = x_0(t - \nu P), \quad t \in [\nu P, (\nu + 1) P], \qquad \nu = 0, \pm 1, \pm 2, \dots .$$

The coefficients C_k may be then expressed for $k = 0, \pm 1, \pm 2, \dots$ in the following way:

$$(2\text{-}5) \qquad C_k = \frac{1}{P} \int_0^P x_0(t)\, \mathrm{e}^{-\mathrm{j}k2\pi t/P}\, \mathrm{d}t = \frac{1}{P} \int_{t_0}^{t_0 + P} x(t)\, \mathrm{e}^{-\mathrm{j}k2\pi t/P}\, \mathrm{d}t,$$

where t_0 is arbitrary.

The form of Fourier series notation used till now applies both for real and for complex functions. In the case of a real function it follows from (2-2) that

$$(2\text{-}6) \qquad C_{-k} = C_k^*, \qquad k = 1, 2, \dots, \quad \mathrm{Im}\, C_0 = 0.$$

Two equivalent notations for the Fourier series are shown below for the case of a real function x ((2-7) and (2-9)):

$$(2\text{-}7) \qquad \xi(t) = a_0 + \sum_{k=1}^{\infty} \left[a_k \cos\left(\frac{k\,.\,2\pi t}{P}\right) + b_k \sin\left(\frac{k\,.\,2\pi t}{P}\right) \right],$$

where

$$(2\text{-}8) \qquad a_0 = \frac{1}{P} \int_0^P x(t)\, \mathrm{d}t,$$

$$a_k = \frac{2}{P} \int_0^P x(t) \cos\left(\frac{k\,.\,2\pi t}{P}\right) \mathrm{d}t,$$

$$b_k = \frac{2}{P} \int_0^P x(t) \sin\left(\frac{k\,.\,2\pi t}{P}\right) \mathrm{d}t, \qquad k = 1, 2, \dots,$$

$$(2\text{-}9) \qquad \xi(t) = A_0 + \sum_{k=1}^{\infty} A_k \cos\left(\frac{k\,.\,2\pi t}{P} + \varphi_k\right),$$

where

$$(2\text{-}10) \qquad A_0 = a_0,$$

$$A_k = \sqrt{(a_k^2 + b_k^2)}, \qquad \cos \varphi_k = a_k/A_k, \quad \sin \varphi_k = -b_k/A_k, \quad k = 1, 2, \dots .$$

Table 1. Conversion of various expressions of the Fourier series coefficients

	$\lvert C_k \rvert$, $\arg C_k$	A_k, φ_k	a_k, b_k
$\lvert C_k \rvert$, $\arg C_k$		$A_0 = C_0$ $A_k = 2\lvert C_k \rvert$, $\varphi_k = \arg C_k$	$a_0 = C_0$ $a_k = 2\,\mathrm{Re}\,C_k$, $b_k = -2\,\mathrm{Im}\,C_k$
A_k, φ_k	$C_0 = A_0$ $\lvert C_k \rvert = \frac{1}{2}A_k$ $\arg C_k = \varphi_k$		$a_0 = A_0$ $a_k = A_k \cos \varphi_k$ $b_k = -A_k \sin \varphi_k$
a_k, b_k	$C_0 = a_0$ $C_k = \frac{1}{2}\sqrt{(a_k^2 + b_k^2)}$ $\cos \varphi_k = \dfrac{a_k}{2\lvert C_k \rvert}$ $\sin \varphi_k = \dfrac{-b_k}{2\lvert C_k \rvert}$ $C_k = \frac{1}{2}a_k - \mathrm{j}\frac{1}{2}b_k$	$A_0 = a_0$ $A_k = \sqrt{(a_k^2 + b_k^2)}$ $\cos \varphi_k = \dfrac{a_k}{A_k}$ $\sin \varphi_k = \dfrac{-b_k}{A_k}$	

x is a real function, $k = 1, 2, 3, \ldots$

Table 1 shows the relationships between coefficients in these three different notations. The complex transcription of a Fourier series in the form (2-1) is, because of its compactness, very convenient and we shall therefore prefer it in the subsequent text.

2.1.2 Physical meaning of the Fourier series

The mathematical fact that a periodic function may be expressed by means of a Fourier series permits us (e.g. by (2-9)) to interpret the periodic motion of some quantity (e.g. a signal) as the superposition of an infinite number of components, namely elementary cosine-shaped curves whose frequency is given just by integer products of a basic frequency $1/P$. The amplitudes of these components are the coefficients A_k and their phases φ_k.

In engineering practice the task of determining the amplitudes A_k and phases φ_k is called *harmonic analysis,* and the elementary curves $x = A_k \cos(2\pi t/P + \varphi_k)$ are called *harmonic components* or just *harmonics.* Even though for real signals it only makes sense to consider real harmonic components with positive frequencies $k \,.\, 2\pi/P$, it may be mathematically more convenient to use a complex expression which includes the negative frequencies. A quantity C_k is then called *the complex amplitude of the k-th harmonic component.*

If the aim of harmonic analysis is to determine coefficients of single harmonics of a periodic function then its opposite operation is *harmonic synthesis* where a resultant function is created from given amplitudes and phases.

The quantities A_k and C_k or their graphic representation with respect to a sequence of k-th harmonics may be called *the spectrum of a signal* expressed by a function x. If a quantity x has a certain physical dimension then, by the definition of C_k, A_k, a_k, b_k, these quantities must possess the same dimension.

2.1.3 Convergence of a Fourier series

We have formally called a function ξ the sum of a Fourier series; it is usually defined as a limit of partial sums

$$\xi(t) = \lim_{n\to\infty} \xi_n(t) = \lim_{n\to\infty} \sum_{k=-n}^{n} C_k \, \mathrm{e}^{\mathrm{j}k2\pi t/P}. \tag{2-11}$$

Now, however, the question arises as to whether this sum exists at all, i.e. whether, and in what sense, the series converges. Furthermore, how does this sum relate to the function x. The general answer to these questions is not simple, nor is it yet fully known. The literature presents a number of criteria for the convergence of Fourier series (see e.g. [59], [84], [39], [50], [89]), but the practical use of these criteria is usually not straightforward. We present here, therefore, some simple sufficient conditions convenient for most of the cases which appear in engineering practice. We first introduce definitions of a piecewise continuous function and a piecewise smooth function.

A function x is called *piecewise continuous* in an interval $[a, b]$ if it is continuous at all points of an interval $[a, b]$ except at a finite number of points t_i, $i = 1, 2, \ldots, n$ where it has discontinuities of the first kind, i.e. where it is not defined or is not continuous but has its own one-sided limits $x(t_i - 0)$ and $x(t_i + 0)$. If $x(t_i - 0) \neq x(t_i + 0)$ we say that a function x has *a step* of magnitude $x(t_i + 0) - x(t_i - 0)$ at the point t_i. If one of the points t_i is a terminal point we require the existence of one-sided limits $x(a + 0)$ or $x(b - 0)$.

A function x is called *piecewise smooth* in an interval $[a, b]$ if both x and $x' = \mathrm{d}x/\mathrm{d}t$ are piecewise continuous in this interval.

(A) Point convergence

We present a theorem on point convergence [84]:

If x is a periodic function which is piecewise smooth on an interval of periodicity $[0, P)$, then the Fourier series of this function

(a) *converges in each point of continuity and moreover we have* $\xi(t) = x(t)$;

(b) *at points of discontinuity* t_i *within an interval of periodicity a Fourier series gives the arithmetic mean of the left limit and the right limit* $\xi(t_i) = \frac{1}{2}[x(t_i + 0) + x(t_i - 0)]$;

(c) *at points* $t = \nu P$, $\nu = 0, \pm 1, \pm 2, \ldots$, *a Fourier series is equal to the arithmetic mean of the one-sided limits at the boundary points of an interval of periodicity*

$$\xi(\nu P) = \frac{1}{2}[x(0+) + x(P - 0)].$$

(B) Uniform convergence

When approximating Fourier series by a partial sum, it is, however, often more convenient to satisfy uniform, rather than point convergence. Again, one can find in the literature mentioned in the beginning of this section a number of criteria for uniform convergence. We list here some simple sufficient conditions [84].

Let a periodic function x *be continuous in a periodicity interval* $[0, P]$ *and let it have there a piecewise continuous derivative* x' *and furthermore let* $x(0) = x(P)$. *Then the Fourier series of a function* x *converge uniformly on an interval* $[0, P]$.

Let a function x *be absolutely integrable in a periodicity interval* $[0, P]$ *and let it be continuous on some interval* (a, b), $0 \leqq a < b \leqq P$, *together with its derivative; then for every* $\delta > 0$ *the Fourier series of a function* x *converges uniformly in the interval* $[a + \delta, b - \delta]$.

(C) Gibbs' phenomenon

When dealing with Fourier series of a function with steps, we must know how the sum of the Fourier series behaves in the neighbourhood of the steps, where the partial sums ξ_n do not converge uniformly. The behaviour of ξ_n as $n \to \infty$ is known as Gibbs' phenomenon. As each function with discontinuities of the first kind may be split into a continuous function and a step function, we illustrate the properties of Gibbs' phenomenon for an elementary discontinuous function with a step of magnitude $2a$ for $t = 0$ defined as follows:

$$x(t) = \begin{cases} a, & 0 < t < P/2, \\ -a, & P/2 < t < P; \end{cases}$$

this function is drawn in Figure 1. We show the behaviour of the partial sums ξ_n of this function in the right neighbourhood of the point $t = 0$.

$$\xi_n(t) = \frac{4a}{\pi} \sum_{k=0}^{(n-1)/2} \frac{1}{2k+1} \sin \frac{(2k+1)\,2\pi t}{P}.$$

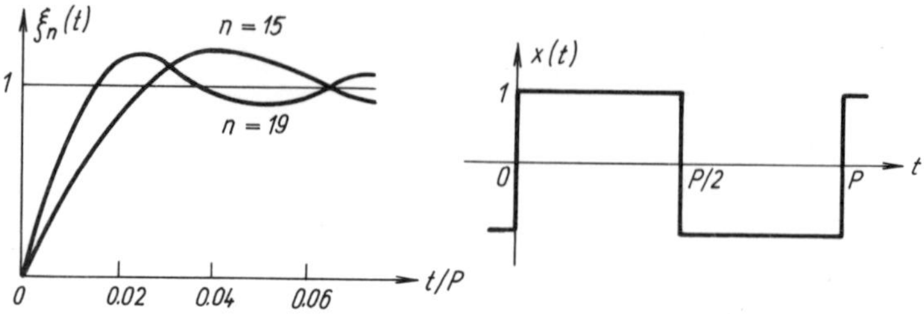

1. Gibbs' phenomenon for the partial sum ξ_n of the Fourier series of the step function x

The curves of the function ξ_n for $n = 15$ and $n = 19$ are shown in Figure 1. It can be demonstrated [39] that ξ_n has extrema in the interval $(0, P/2)$ at points

$$t_{\nu,n} = \frac{\nu P}{2(n+1)}, \qquad \nu = 1, 2, \ldots, n,$$

which has maxima for odd ν and minima for even ν, of value

$$\text{(2-12)} \qquad \xi_n(t_{\nu,n}) = \frac{4a}{\pi} \sum_{k=0}^{(n-1)/2} \frac{1}{2k+1} \sin \nu \frac{(2k+1)\pi}{n+1}.$$

For sufficiently large n, the extrema can be approximated by their limit values which are

$$\text{(2-13)} \qquad \lim_{n\to\infty} \xi_n(t_{\nu,n}) = \frac{2a}{\pi} \operatorname{Si}(\nu\pi).$$

The extrema broaden with growing n, and for sufficiently large n do not go below the given limiting values. The difference between the largest extreme value and the value of the step $x(0+) - x(-0) = 2a$, expressed in per cent of a step magnitude, is 8.95 % for the first maximum, 3.31 % for the second maximum and -4.89 % for the first minimum. More detailed discussion on Gibbs' phenomenon is given in [39] and some of its other properties are given in [8].

(D) Properties of the coefficients C_k

When calculating the coefficients, it is advantageous to know their behaviour as $k \to \infty$. This is determined by the following theorem [83], [84]:

Let x be a periodic function with an interval of periodicity $[0, P)$, *continuous together with its derivatives up to order* $\varkappa - 1$, $\varkappa \geqq 1$. *Let its derivative of order $\varkappa$ be piecewise continuous in the interval. Then* *there exists $M > 0$ such that, for Fourier series coefficients of a function x, we have*

$$(2\text{-}14) \qquad |C_k| \leqq \frac{M}{|k|^{\varkappa+1}}.$$

Thus, for example, a continuous periodic function with steps in a first-order derivative has coefficients which meet the following condition

$$|C_k| \leqq \frac{M}{k^2}.$$

It can also be proved [59] that the coefficients C_k of a periodic function x, which has steps in a periodicity interval and, at the same time, has first order derivatives at all other points, satisfy the condition

$$|C_k| \leqq \frac{M}{|k|}.$$

2.1.4 Convergence in mean

At a point convergence of a partial sum ξ_n as well as at a uniform one, we have evaluated its closeness to the function x as the difference between a limit of partial sums and the given function x either at single points or at all points of a periodicity interval. The theory of Fourier series, however, permits us to investigate this closeness by a more general procedure which we call *convergence in mean.* Although this procedure of evaluating the mutual closeness of two curves lacks the advantages of a simple geometric image of the distance between two points, it does have the advantage of applying to a substantially wider class of functions x which can be expanded into Fourier series converging in this sense.

The theory of Fourier series based upon convergence in mean applies to functions from a space $L_2(0, P)$ [50], [59], i.e. for square-integrable functions. These are functions for which the Lebesgue integral

$$\int_0^P |x(t)|^2 \, dt$$

converges. From a practical point of view it is convenient that all piecewise continuous functions belong to $L_2(0, P)$ (see § 2.1.3).

We now present a definition of convergence in mean.

If ξ_n, $n = 1, 2, \ldots$, is a sequence of functions from $L_2(0, P)$, then we say that ξ_n converges to x in quadratic mean if

$$\lim_{n \to \infty} \int_0^P |\xi_n(t) - x(t)|^2 \, dt = 0.$$

That is why this convergence is also called *a convergence in the sense of quadratic deviation.*

We can now, at last, state a theorem on Fourier series:

Let x be a function from a class $L_2(0, P)$. Then its Fourier series (2-1) *exists with coefficients given by* (2-2) *and which has the following properties:*

1. *A sequence of partial sums ξ_n converges to x in the mean,* i.e.

$$\lim_{n\to\infty} \int_0^P |\xi_n(t) - x(t)|^2 \, dt = 0. \tag{2-17}$$

2. *Between a function x and the coefficients of its Fourier series, Parseval's equality holds*

$$\frac{1}{P} \int_0^P |x(t)|^2 \, dt = \sum_{k=-\infty}^{\infty} |C_k|^2. \tag{2-18}$$

3. *The coefficients of a Fourier series converge to zero:*

$$\lim_{k\to\infty} C_k = 0. \tag{2-19}$$

On the basis of convergence in mean of the Fourier series the following very important theorem may be formulated, that is in practice used to approximate a function by a trigonometric polynomial:

If a function x from a class $L_2(0, P)$ is to be approximated by an n-th order trigonometric polynomial (there exist also other criteria of approximation by a trigonometric polynomial, see e.g. [52])

$$\eta_n(t) = \sum_{k=-n}^{n} K_k \, e^{jk2\pi t/P}$$

in the sense of minimum quadratic deviation, i.e. so that an integral

$$\int_0^P |x(t) - \xi_n(t)|^2 \, dt = \varepsilon_n^3$$

would be minimised, then of all the polynomials of n-th order the polynomial with such properties is just a partial sum of a Fourier series of function x, i.e.

$$\eta_n(t) = \xi_n(t),$$

and hence

$$K_k = C_k, \qquad k = 0, \pm 1, \pm 2, \ldots, \pm n.$$

For the quadratic deviation Bessel's inequality applies

$$\varepsilon_n^2 = \int_0^P |x(t)|^2 \, dt - P \sum_{k=-n}^{n} |C_k|^2 \geqq 0.$$

All the expressions presented in this paragraph are special cases of much more general ones which hold in the theory of general orthogonal series in Hilbert spaces (see e.g. [59]).

2.2 Approximation by a trigonometric polynomial with a minimum quadratic deviation at given points

In practice, a function x of a variable t, $t \in [0, P)$ is not always known at all points but we may know just its function values x_i at discrete points. We show now how, in these circumstances, one can construct a trigonometric polynomial through the values x_i similar to that of a previous paragraph.

2.2.1 Formulation of the problem and its solution

Let us assume that a function x (real or complex) of a variable t, $t \in [0, P)$ is known at N equidistant points $x(iT) = x_i$, $i = 0, 1, \ldots, N-1$, $T = P/N$, and we are looking for a trigonometric polynomial

$$u_M(t) = \sum_{k=-M}^{M} D_k \, e^{jk2\pi t/P} \tag{2-21}$$

such that the sum of the squares of the absolute values of deviations $x_i - u_M(iT)$ would be minimal, i.e. the sum

$$\sum_{i=0}^{N-1} |x_i - u_M(iT)|^2 = \delta_M^2 \tag{2-22}$$

would be minimal. If we solve the task as an extremum of the function δ_M^2 of variables $D_\varkappa$, $\varkappa = 0, \pm 1, \pm 2, \ldots, \pm M$, we arrive at a set of $2M + 1$ so-called *normal equations* for $2M + 1$ coefficients D_k

$$\sum_{i=0}^{N-1} e^{-ji\varkappa 2\pi/N} \Big(x_i - \sum_{k=-M}^{M} D_k \, e^{jik2\pi/N}\Big) = 0, \qquad \varkappa = 0, \pm 1, \ldots, \pm M,$$

which may be adjusted to a set

$$\sum_{i=0}^{N-1} x_i \, e^{-ji\varkappa 2\pi/N} = \sum_{i=0}^{N-1} \sum_{k=-M}^{M} D_k \, e^{ji(k-\varkappa)2\pi/N}. \tag{2-23}$$

If we apply the orthogonal properties of the sequence $e^{jik2\pi/N}$ i.e.

$$\sum_{i=0}^{N-1} e^{ji(k-\varkappa)2\pi/N} = \begin{cases} N, & k = \varkappa + \nu N, \\ 0, & k \neq \varkappa + \nu N, \end{cases} \qquad \nu = 0, \pm 1, \pm 2, \ldots,$$

we obtain the set of equations

$$\sum_{i=0}^{N-1} x_i \, e^{-ji\varkappa 2\pi/N} = N \sum_{\varkappa+\nu N=-M}^{M} D_{\varkappa+\nu N}, \qquad \nu = 0, \pm 1, \pm 2, \ldots, \quad \varkappa = 0, \pm 1, \ldots, \pm M. \tag{2-24a}$$

In order that each of the $2M + 1$ equations contains just one D_k, it must be true that $M < N/2$. Then the following expression for D_k applies:

$$\text{(2-24b)} \quad D_k = \frac{1}{N} \sum_{i=0}^{N-1} x_i \, \mathrm{e}^{-\mathrm{j}ik2\pi/N}, \qquad k = 0, \pm 1, \pm 2, \ldots, \pm M.$$

This means that, for even N, $M \leqq N/2 - 1$ and for odd N, $M \leqq (N - 1)/2$.

Define the sum of squares of absolute values of deviations δ_M^2 by the relation

$$\delta_M^2 = \sum_{i=0}^{N-1} |x_i - u_i|^2 = \sum_{i=0}^{N-1} (x_i - u_i)(x_i^* - u_i^*).$$

Then the following relations, which can be proved by direct calculation:

$$\sum_{i=0}^{N-1} u_i x_i^* = N \sum_{k=-M}^{M} |D_k|^2$$

and

$$\sum_{i=0}^{N-1} |u_i|^2 = N \sum_{k=-M}^{M} |D_k|^2,$$

can be used to derive the result

$$\text{(2-25a)} \quad \delta_M^2 = \sum_{i=0}^{N-1} |x_i|^2 - N \sum_{k=-M}^{M} |D_k|^2 \geqq 0.$$

The quantity δ_M^2 is non-negative; for $M = 2, 3, \ldots, N/2 - 1$ in the case where N is even or for $M = 2, 3, \ldots, (N - 1)/2$ in the case where N is odd we have the inequality

$$\text{(2-25b)} \quad \delta_M^2 \leqq \delta_{M-1}^2.$$

Given $x_i = x(iT)$, a polynomial of the required properties can be set up for every $M < N/2$. For constant N, the quadratic deviation decreases with growing M. At the same time a polynomial u_M of order $M \geqq 1$ has coefficients D_k for $k = 0, 1, \ldots, M - 1$ identical to the corresponding coefficients of a polynomial u_{M-1} of order $M - 1$. A drawback of this method of approximation is that it is not in general possible to state anything about the behaviour of polynomials u_M between points iT.

2.2.2 Relation to interpolation

(A) Odd N

If we choose for a given N the highest possible order $M = N/2 - 1$ so that $N = 2M + 1$, then after substituting for D_k in (2-24) we obtain from (2-25a) the relation

$$\text{(2-26)} \quad N \sum_{k=-N/2}^{N/2} |D_k|^2 = \sum_{i=0}^{N-1} |x_i|^2,$$

which we shall call *Parseval's equality.* From this it follows that, in the case $N = 2M + 1$, the quadratic deviation δ_M^2 is equal to zero. Single deviations $x_i - u_{N/2-1}(iT)$ are, therefore, equal to zero too and it is clear that the trigonometric polynomial $u_{N/2-1}$ interpolates the values x_i, $i = 0, 1, \ldots, N-1$.

(B) Even N

On the basis of the preceding consideration, it would be possible to achieve an interpolation for odd N only. Let us consider now the case where $N = 2M$. Then the set (2-24) becomes a set

$$\text{(2-27a)} \quad D_k = N^{-1} \sum_{i=0}^{N-1} x_i \, \mathrm{e}^{-\mathrm{j}ik2\pi/N}, \qquad k = 0, \pm 1, \ldots, \pm(N/2-1),$$

and an equation valid for $k = \pm N/2$. This equation allows us to determine only a sum of coefficients which we write as $D'_{N/2} + D'_{-N/2}$:

$$\text{(2-27b)} \quad D'_{N/2} + D'_{-N/2} = N^{-1} \sum_{i=0}^{N-1} x_i(-1)^i.$$

As there are available $2M$ equations altogether, we can determine only $2M$ coefficients. We, therefore, adjust the term of the highest power in the trigonometric polynomial (2-21) so that the polynomial has only $2M$ coefficients:

$$\text{(2-28)} \quad u_{N/2}(t) = \sum_{k=-(N/2-1)}^{N/2-1} D_k \, \mathrm{e}^{\mathrm{j}k2\pi t/P} + \frac{1}{2}(D'_{N/2} + D'_{-N/2})(\mathrm{e}^{\mathrm{j}N\pi t/P} + \mathrm{e}^{-\mathrm{j}N\pi t/P}).$$

We shall further define

$$\text{(2-29)} \quad D_{N/2} = D_{-N/2} = \frac{1}{2}(D'_{N/2} + D'_{-N/2}) = \frac{1}{2N} \sum_{i=0}^{N-1} x_i(-1)^i,$$

which permits us to write a trigonometric polynomial in the form

$$\text{(2-30)} \quad u_{N/2}(t) = \sum_{k=-N/2}^{N/2} D_k \, \mathrm{e}^{\mathrm{j}k2\pi t/P}.$$

We show that in this case an interpolation also occurs. For $t = iT$ we get from (2-28) after substituting D_k from (2-24b) and $D_{N/2}$ from (2-29)

$$u_{N/2}(iT) = \frac{1}{N} \sum_{\nu=0}^{N-1} x_\nu \sum_{k=-(N/2-1)}^{N/2-1} \mathrm{e}^{\mathrm{j}(i-\nu)k2\pi/N} + \frac{1}{N}(-1)^i \sum_{\nu=0}^{N-1} (-1)^\nu x_\nu.$$

Using the sum

$$\sum_{k=-(N/2-1)}^{N/2-1} e^{jmk2\pi/N} = \begin{cases} N-1, & m = 0, \\ -(-1)^m, & m \neq 0, \end{cases}$$

we get for $i = 0, 1, \ldots, N-1$

$$u_{N/2}(iT) = \frac{1}{N}\Big[-(-1)^i \sum_{\substack{\nu=0 \\ \nu \neq i}}^{N-1} x_\nu(-1)^\nu + (N-1)\,x_i +$$

$$+ (-1)^i \sum_{\nu=0}^{N} (-1)^\nu x_\nu\Big] = x_i.$$

(C) Summary

Every trigonometric polynomial (2-21) of order

$$M \leqq \lceil (N-1)/2 \rceil,$$

(where $\lceil\ \rceil$ denotes the integer part of a number) which approximates values x_i at points iT, $i = 0, 1, \ldots, N-1$, in the sense of minimum deviation, has coefficients D_k determined by the expression (2-24b) for $k = 0, \pm 1, \pm 2, \ldots, \pm M$. For odd N, a polynomial of degree $M = (N-1)/2$ is an interpolation polynomial. For even N, an interpolation polynomial of degree $M = N/2$ is given by relation (2-30). Its coefficients are determined by relation (2-24b) for $k = 0, \pm 1, \pm 2, \ldots, N/2 - 1$ and by relation (2-29) for $k = N/2$.

Another way of expressing interpolation polynomials is presented in § 5.3.1 where it is also demonstrated that relation (2-24b) closely relates to the discrete Fourier transform.

2.2.3 Relation to coefficients of the Fourier series

If we assume that $x_i = x(iT)$, $i = 0, 1, \ldots, N-1$ are values of a function x with a Fourier series

$$x(t) = \sum_{\varkappa=-\infty}^{\infty} C_\varkappa e^{i\varkappa 2\pi t/P},$$

then there holds for them the relation

$$x_i = \sum_{\varkappa=-\infty}^{\infty} C_\varkappa e^{ji\varkappa 2\pi/N}, \qquad i = 0, 1, \ldots, N-1. \tag{2-31}$$

If we now substitute this expression into (2-24b), we get for D_k the relation

$$D_k = N^{-1} \sum_{i=0}^{N-1} \sum_{\varkappa=-\infty}^{\infty} C_k e^{ji(\varkappa-k)2\pi/N}.$$

Using the relation

$$\sum_{i=0}^{N-1} e^{jim2\pi/N} = \begin{cases} 0, & m \neq \nu N, \\ N, & m = \nu N, \end{cases} \qquad \nu = 0, \pm 1, \pm 2, \dots,$$

we get for D_k the expression

(2-32) $$D_k = \sum_{\nu=-\infty}^{\infty} C_{k+\nu N}$$

giving a relationship between the coefficients D_k of a trigonometric polynomial and the coefficients of a Fourier series. This relationship will be discussed further in § 5.1.1.

2.3 Distributions

In many applications, e.g. for discrete signal processing and in the theory of systems, it is reasonable to use the δ distribution. As the δ distribution can be used also for expressing the relationship of the discrete Fourier transform to the Fourier series and the Fourier transform, we describe some of its properties here.

2.3.1 Definition of a distribution

The δ distribution (*generalised delta function, Dirac function, delta function*) is a singular distribution defined as a linear and continuous functional $\langle \delta, \varphi \rangle = \varphi(0)$ over a space of admissible functions φ [42], [9]. From the theory of distributions (generalised functions) it follows that any function of class L_2 can be expressed as a regular distribution. For our purpose it will be more advantageous to introduce the δ distribution as the generalised limit of a sequence of functions. This concept corresponds to a common physical notion of the δ distribution as the limiting case of a narrow impulse with unit area.

We introduce first a so-called δ sequence by the definition [42]:

Let a sequence of functions (or distributions) q_n have the following properties: There exists an $M > 0$ so that for all a, b for which $|a| \leqq M$, $|b| \leqq M$, there holds:

1. $\left| \int_a^b q_n(t) \, dt \right| < L$,

where L is a real constant independent of a, b;

2. If $a \neq 0$, $b \neq 0$, then

$$\lim_{n \to \infty} \int_a^b q_n(t) \, dt = \begin{cases} 0, & a < b < 0, \quad 0 < a < b, \\ 1, & a < 0 < b. \end{cases}$$

This sequence we then call a δ *sequence.* We define the δ distribution as follows:

The δ distribution is a limit of the δ sequence. In order to simplify notation and avoid writing out the δ distribution as a functional, we shall from now on write this distribution just like a function. We may then write

$$\lim_{n\to\infty} q_n(t) = \delta(t) \tag{2-33}$$

and this limit will be called the δ *distribution at the point* $t = 0$.

We now present examples of δ sequences:

1. $q_n(t) = \begin{cases} n, & 0 < t < 1/n, \\ 0, & t < 0, \quad t > 1/n; \end{cases}$

2. $q_n(t) = \begin{cases} n^2(t + 1/n), & -1/n \leqq t \leqq 0, \\ -n^2(t - 1/n), & 0 \leqq t \leqq 1/n; \end{cases}$

3. $q_n(t) = \dfrac{1}{2}\sqrt{\left(\dfrac{n}{\pi}\right)}\, e^{-nt^2/2};$

4. $q_n(t) = \dfrac{1}{\pi}\dfrac{1/n}{t^2 + 1/n^2};$

5. $q_n(t) = \dfrac{1}{\pi}\dfrac{\sin nt}{t}.$

Furthermore we present some rules for operations with the δ distribution which permits us to treat it as a function. We mention, however, that these rules apply only to the applications shown later in the book and warn readers against extending their validity to other cases. Proofs of properties of the δ distribution from which the rules result can be found e.g. in [42], [77], [7], [9].

2.3.2 Rules for operations with δ distributions

1. *A shifted δ distribution,* i.e. a δ distribution at a point t_0, will be written as $\delta(t - t_0)$, $t_0 \lessgtr 0$, while for $t_0 = 0$ we have $\delta(t - t_0) = \delta(t)$.

2. If we change a variable t in the δ distribution to a variable at, a real, then

$$\delta(at) = \frac{1}{|a|}\,\delta(t). \tag{2-34}$$

3. For each t_0 where a function x is continuous there holds for the product of the function x with the δ distribution:

$$x(t)\,\delta(t - t_0) = x(t_0)\,\delta(t - t_0). \tag{2-35}$$

Thus the product of the δ distribution at a point t_0 with a function is equal to the δ distribution multiplied by the value of the function at a point t_0.

4. For an integral of the product of a function x and the δ distribution the following is true:

$$\text{(2-36)} \quad \int_a^b \delta(t - t_0)\, x(t)\, \mathrm{d}t = \begin{cases} 0, & t_0 > b, \quad t_0 < a, \\ & -\infty \leqq a < b \leqq \infty, \\ x(t_0), & a < t_0 < b. \end{cases}$$

The integral of the product of the δ distribution and a function equals the value of the function at the location of the distribution.

5. For the convolution of a function or distribution x with the δ distribution defined as

$$x(t) * \delta(t - t_0) = \int_{-\infty}^{\infty} x(t - \tau)\, \delta(\tau - t_0)\, \mathrm{d}\tau,$$

then

$$\text{(2-37)} \quad x(t) * \delta(t - t_0) = x(t - t_0).$$

The convolution of a function x with the δ distribution at a point t_0 is equal to a translation of the function x by a length t_0 [76]. As a result of this it follows that

$$\text{(2-38)} \quad \delta(t - t_1) * \delta(t - t_2) = \delta(t - (t_1 + t_2)).$$

2.3.3 Fourier series of distributions

The theory of distributions allows us to introduce periodic distributions [42], [7] as well. They can also be expressed by a Fourier series, the form of the series being the same as the expression of the Fourier series by the functions (2-1) including relation (2-2) for the coefficients [7]. In addition, each trigonometric series is a Fourier series of some periodic distribution and converges in a generalised sense, if there exists a $\varkappa \geqq 0$ so that $|C_k|/k^\varkappa \to 0$ for $k \to \infty$.

A periodic distribution δ with period P is defined to be a distribution of the form

$$\sum_{i=-\infty}^{\infty} \delta(t - iP).$$

This distribution can be split into a Fourier series [7]

$$\sum_{k=-\infty}^{\infty} C_k\, \mathrm{e}^{\mathrm{j}k2\pi t/P}, \qquad C_k = \frac{1}{P},$$

converging in a generalised sense; hence

$$\text{(2-39)} \quad \sum_{i=-\infty}^{\infty} \delta(t - iP) = \frac{1}{P} \sum_{k=-\infty}^{\infty} \mathrm{e}^{\mathrm{j}k2\pi t/P}.$$

2.4 Fourier transforms

2.4.1 Definition of a Fourier transform

We define the direct Fourier transform

$$(2\text{-}40)\qquad X(\omega) = \int_{-\infty}^{\infty} x(t)\,\mathrm{e}^{-\mathrm{j}\omega t}\,\mathrm{d}t$$

and a corresponding relation

$$(2\text{-}41)\qquad x(t) = \frac{1}{2\pi}\int_{-\infty}^{\infty} X(\omega)\,\mathrm{e}^{\mathrm{j}\omega t}\,\mathrm{d}\omega$$

for the inverse one.

We shall call a complex function X of a real variable ω defined by relation (2-40) *the Fourier transform or image* of a complex or real function x. Symbolically we shall denote the Fourier transform as follows:

$$(2\text{-}42)\qquad X(\omega) = \mathscr{F}\{x(t)\}, \qquad x(t) = \mathscr{F}^{-1}\{X(\omega)\}$$

or briefly

$$x(t) \triangleq X(\omega).$$

Some conditions of convergence of the definition integrals may be found e.g. in [16], [19], [67], [68], [39], [59], [8]. For our purpose we content ourselves with this sufficient condition [85]:

If there exists a Lebesgue integral

$$\int_{-\infty}^{\infty} |x(t)|\,\mathrm{d}t$$

and the function x is piecewise smooth, then its Fourier transform X exists.

For the Fourier transform of functions which satisfy this sufficient condition the following applies:

$$\lim_{\omega\to\infty} X(\omega) = 0.$$

We present here just one other result which is important for numerical computation of the Fourier transform:

For the Fourier transform X of a function x which is continuous including its derivatives up to order $k-1$, $k \geqq 1$, and whose derivative of k-th order is piecewise continuous, there exists an $M > 0$ such that

$$(2\text{-}43)\qquad |X(\omega)| \leqq M|\omega|^{-k-1}.$$

It is obvious that this is analogous to a theorem from § 2.1.3 D concerned with properties of coefficients of the Fourier series. It is necessary to mention, however, that the Fourier transforms of distributions do not generally possess this property.

2.4.2 Relationship of Fourier transforms to Fourier series

Since for finite functions the conditions for the existence of a Fourier transform are the same as for convergence of the Fourier series of a periodised function, we may determine the Fourier series coefficients for such a function

$$C_k = \frac{1}{P}\int_0^P x(t)\,\mathrm{e}^{-\mathrm{j}k2\pi t/P}\,\mathrm{d}t, \qquad k = 0, \pm 1, \dots,$$

as well as the Fourier transform by (2-40)

$$X(\omega) = \int_0^P X(t)\,\mathrm{e}^{-\mathrm{j}\omega t}\,\mathrm{d}t.$$

Comparison of these relations allows us to make this conclusion:

The coefficients C_k of a Fourier series of a finite function x are equal to the values of a Fourier transform X of this function for $\omega = k\,.\,2\pi/P$, $k = 0, \pm 1, \pm 2, \dots$, divided by the period:

$$C_k = \frac{1}{P} X\left(\frac{k\,.\,2\pi}{P}\right). \tag{2-44}$$

2.4.3 Some properties of Fourier transforms

We present here some basic properties of Fourier transforms. Their proofs, as well as other properties, can be found e.g. in [6], [17], [16], [59], [67], [68], [70], [85].

1. Shifting of a function and multiplication by an exponential

$$x(t - t_0) \mathrel{\hat{=}} X(\omega)\,\mathrm{e}^{-\mathrm{j}\omega t_0}, \qquad t_0 \text{ real;} \tag{2-45}$$

$$X(\omega - \omega_0) \mathrel{\hat{=}} x(t)\,\mathrm{e}^{\mathrm{j}\omega_0 t}, \qquad \omega_0 \text{ real.} \tag{2-46}$$

2. Transform of a variable

$$x(at) \mathrel{\hat{=}} \frac{1}{|a|} X\left(\frac{\omega}{a}\right), \qquad a \text{ real, } a \neq 0. \tag{2-47}$$

3. Convolution and product of functions

$$x(t) * y(t) \mathrel{\hat{=}} X(\omega)\,Y(\omega); \tag{2-48}$$

$$x(t)\,y(t) \mathrel{\hat{=}} \frac{1}{2\pi} X(\omega) * Y(\omega), \tag{2-49}$$

where the symbol $*$ means a convolution,

$$x(t) * y(t) = \int_{-\infty}^{\infty} x(t - \tau)\,y(\tau)\,\mathrm{d}t = \int_{-\infty}^{\infty} x(\tau)\,x(t - \tau)\,\mathrm{d}\tau, \tag{2-50}$$

and

$$x(t) \mathrel{\hat{=}} X(\omega), \qquad y(t) \mathrel{\hat{=}} Y(\omega).$$

2.4.4 Fourier transforms of distributions

Fourier transforms can also be defined for distributions, again just as a functional over a space of suitable functions [42], [7]. It can be proved that the properties 1, 2 and 3 from § 2.4.3 hold also for distributions. We present here some values of the Fourier transforms where the δ distribution appears.

1. Fourier transform of the δ distribution

(2-51) $\delta(t) \mathrel{\hat{=}} 1;$

(2-52) $c\delta(t) \mathrel{\hat{=}} c.$

2. Fourier transform of the δ distribution at a point $t = t_0$

(2-53) $\delta(t - t_0) \mathrel{\hat{=}} \mathrm{e}^{-\mathrm{j}\omega t_0}, \qquad t_0$ real.

3. Fourier transform of a constant

(2-54) $1 \mathrel{\hat{=}} 2\pi\delta(\omega);$

(2-55) $K \mathrel{\hat{=}} 2\pi K\delta(\omega).$

4. Fourier transform of a complex exponential function

(2-56) $\mathrm{e}^{\mathrm{j}\omega_0 t} \mathrel{\hat{=}} 2\pi\delta(\omega - \omega_0), \qquad \omega_0$ real.

5. Fourier transform of a periodic δ distribution. Using (2-53) we get

(2-57) $$\sum_{i=-\infty}^{\infty} \delta(t - iP) \mathrel{\hat{=}} \sum_{\nu=-\infty}^{\infty} \mathrm{e}^{-\mathrm{j}\omega iP}.$$

If we start from an expression of this periodic distribution by a Fourier series, i.e. from relation (2-39), then on applying the Fourier transform to the right side of this relation we obtain a value

(2-58) $$\frac{1}{P}\sum_{k=-\infty}^{\infty} \mathrm{e}^{-\mathrm{j}k2\pi t/P} \mathrel{\hat{=}} \frac{2\pi}{P}\sum_{k=-\infty}^{\infty} \delta\left(\omega - \frac{k\,.\,2\pi}{P}\right).$$

As a result this identity holds

(2-59) $$\sum_{i=-\infty}^{\infty} \mathrm{e}^{-\mathrm{j}\omega iP} = \frac{2\pi}{P}\sum_{k=-\infty}^{\infty} \delta\left(\omega - \frac{k\,.\,2\pi}{P}\right).$$

6. Fourier transform of a periodic function

A periodic function x_p can be expressed as an infinite sum of shifted functions $x(t - \nu P)$ by means of convolution with a periodic distribution δ;

(2-60) $$x_p(t) = \sum_{\nu=-\infty}^{\infty} x(t - \nu P) = x(t) * \sum_{\nu=-\infty}^{\infty} \delta(t - \nu P).$$

If we use for the Fourier transform of a function x, relation (2-48) for the image of a convolution and relation (2-57), we get

(2-61) $$x_p(t) \mathrel{\hat{=}} X(\omega) \sum_{\nu=-\infty}^{\infty} \mathrm{e}^{-\mathrm{j}\omega\nu P}.$$

If we now express the infinite sum by means of identity (2-59), we obtain an expression

$$(2\text{-}62)\qquad x_p(t) \triangleq \frac{2\pi}{P} X(\omega) \sum_{k=-\infty}^{\infty} \delta\left(\omega - \frac{k \, . \, 2\pi}{P}\right),$$

which can be further adjusted, by rule (2-35) on the product of a function with the δ distribution, into the form

$$(2\text{-}63)\qquad x_p(t) \triangleq \frac{2\pi}{P} \sum_{k=-\infty}^{\infty} X\left(\frac{k \, . \, 2\pi}{P}\right) \delta\left(\omega - \frac{k \, . \, 2\pi}{P}\right).$$

If we now make an inverse transform using (2-56), we get the expression

$$(2\text{-}64)\qquad x_p(t) = \frac{1}{P} \sum_{k=-\infty}^{\infty} X\left(\frac{k \, . \, 2\pi}{P}\right) \mathrm{e}^{\mathrm{j}k2\pi t/P},$$

which is a trigonometric series and by § 2.3.3 even a Fourier series, provided that $X(k \, . \, 2\pi/P)$ satisfies the condition presented in that section.

Expression (2-63) then gives the Fourier transform of a Fourier series. If we express a function x_p by a Fourier series

$$(2\text{-}65)\qquad x_p(t) = \sum_{k=-\infty}^{\infty} C_k \, \mathrm{e}^{\mathrm{j}k2\pi t/P}$$

and carry out its Fourier transform using (2-56), we get

$$(2\text{-}66)\qquad x_p(t) = \sum_{k=-\infty}^{\infty} 2\pi C_k \delta\left(\omega - \frac{k \, . \, 2\pi}{P}\right).$$

In the case where x is a finite function (2-44) holds and (2-66) is identical to (2-63).

2.4.5 Hilbert transforms and their relation to Fourier transforms

Hilbert transforms (in a hyperbolic form) are usually defined [38] by a relation

$$(2\text{-}67)\qquad y(t) = \frac{1}{\pi} \int_{-\infty}^{\infty} \frac{x(\tau)}{\tau - t} \, \mathrm{d}\tau = \mathscr{H}\{x(t)\},$$

where the integral is understood in the sense of the principal value

$$y(t) = \lim_{\substack{A \to \infty \\ \varepsilon \to 0}} \left[\int_{-A}^{t-\varepsilon} \frac{x(\tau)}{\tau - t} \, \mathrm{d}\tau + \int_{t+\varepsilon}^{A} \frac{x(\tau)}{\tau - t} \, \mathrm{d}\tau\right].$$

A sufficient condition for the existence of an image is that the function x belongs to $L_2(-\infty, \infty)$.

The Hilbert transform can be defined also for distributions [7]. If we start from a distribution $1/(\pi t)$ for whose Fourier transform

$$(2\text{-}68)\qquad \frac{1}{\pi t} \triangleq -\mathrm{j} \operatorname{sign} \omega,$$

where

$$\text{(2-69)} \qquad \operatorname{sign}\omega = \begin{cases} 1, & \omega > 0, \\ 0, & \omega = 0, \\ -1, & \omega < 0, \end{cases}$$

we can write the Hilbert transform as a convolution

$$\text{(2-70)} \qquad y(t) = -(\pi t)^{-1} * x(t).$$

A. If Fourier transforms of functions $x(t) \triangleq X(\omega)$ and $y(t) \triangleq Y(\omega)$ exist, then from the theorem on the image of convolutions (2-48) follows the following relation between the Fourier image of a function and the image of its Hilbert transform:

$$\text{(2-71)} \qquad Y(\omega) = \mathrm{j}X(\omega) \operatorname{sign}\omega.$$

B. On that basis we may determine the Hilbert transform of the function $x(t) = \mathrm{e}^{\mathrm{j}\omega_0 t}$, $\omega_0 \neq 0$. We obtain

$$Y(\omega) = \mathrm{j}\, 2\pi\delta(\omega - \omega_0) \operatorname{sign}\omega = \mathrm{j}\, 2\pi\delta(\omega - \omega_0) \operatorname{sign}\omega_0$$

and from that the result

$$\text{(2-72)} \qquad y(t) = \mathscr{H}\{\mathrm{e}^{\mathrm{j}\omega_0 t}\} = \mathrm{j}\, \mathrm{e}^{\mathrm{j}\omega_0 t} \operatorname{sign}\omega_0,$$

which we use to introduce the discrete Hilbert transform.

C. If we apply property (2-71) to the Fourier transform of a periodic function expressed by a Fourier series

$$\text{(2-73)} \qquad x_p(t) \triangleq \sum_{k=-\infty}^{\infty} 2\pi C_k \delta\left(\omega - \frac{k \,.\, 2\pi}{P}\right),$$

we get

$$y_p(t) \triangleq \mathrm{j} \operatorname{sign}\omega \sum_{k=-\infty}^{\infty} 2\pi C_k \delta\left(\omega - \frac{k \,.\, 2\pi}{P}\right) =$$

$$= \mathrm{j} \,.\, 2\pi \sum_{k=-\infty}^{\infty} C_k \delta\left(\omega - \frac{k \,.\, 2\pi}{P}\right) \operatorname{sign} k.$$

By the inverse transform we get the expression

$$\text{(2-74)} \qquad y_p(t) = \mathrm{j} \sum_{k=-\infty}^{\infty} C_k \, \mathrm{e}^{\mathrm{j}k 2\pi t/P} \operatorname{sign} k$$

which defines the conjugate Fourier series [50], [59].

3. PRACTICAL METHODS OF COMPUTING FOURIER TRANSFORMS

3.1 Problems of computing Fourier transforms

3.1.1 Basic considerations concerning the computation of Fourier transforms

As the relations for computing coefficients of Fourier series in complex form and the defining relations for direct and inverse Fourier transforms are very similar, we consider further, for simplicity, just the computation of an integral

$$F(y) = \int_A^B f(x)\, e^{jyx}\, dx, \qquad -\infty \leqq A < y < B \leqq \infty, \tag{3-1}$$

which we call the *Fourier transform.*

For computing Fourier transforms and particularly for the purpose of harmonic analysis a large number of methods has been worked out. This resulted from the fact that Fourier transforms are used in a number of areas, so that there are very different requirements for methods of computation. There would be no reason to present here a review of methods developed in the past, as developments in computers have established a preference first of all for methods based upon discrete Fourier transforms. Progress in these methods followed the development of very effective algorithms for computing discrete Fourier transforms. We limit ourselves, therefore, in this chapter to presenting some methods that can be well integrated with the use of discrete Fourier transforms or which somehow relate to them. At the same time, we would like to show what procedures in computing Fourier transforms might be considered of practical use and effectiveness.

In no case do we wish to encourage the reader to think that methods not using the discrete Fourier transforms are inefficient. A careful choice of the method of computing a Fourier transform should be made for the solution of any task and especially with tasks of a special nature. Unfortunately, there exists no evaluation nor a general review of methods of computing Fourier transforms. Some recent methods are described in [34], [57] and [58]. Older methods are very thoroughly collected in [76].

Let us consider now the case where a function f_A is constant or exponentially decreasing:

(3-5) $$f_A(x) = f_A(x_A)\,e^{-\alpha(x-x_A)}, \qquad \alpha \geqq 0, \quad x_A > 0$$

Then there holds

$$F_A(y) = f(x_A)\,(A + jB),$$

where, for $\alpha > 0$,

(3-6) $$A = e^{\alpha x_A} \int_{x_A}^{\infty} e^{-\alpha x} \cos yx \, dx = \frac{\alpha \cos x_A y - y \sin x_A y}{\alpha^2 + y^2},$$

(3-7) $$B = e^{\alpha x_A} \int_{x_A}^{\infty} e^{-\alpha x} \sin yx \, dx = \frac{\alpha \sin x_A y + y \cos x_A y}{\alpha^2 + y^2}$$

and for $\alpha = 0$

(3-8) $$A = \pi\delta(y) - y^{-1} \sin x_A y,$$

(3-9) $$B = y^{-1} \cos x_A y.$$

As another example let us consider a function f_A which is hyperbolically decreasing:

(3-10) $$f_A(x) = f_A(x_A)\,(x_A/x)^n, \qquad n \geqq 1, \quad x_A > 0.$$

For $F_A(y)$ we then get

$$F_A(y) = f_A(x_A)\, x_A^n (C_n + jS_n),$$

where

(3-11) $$C_n = y^{n-1} \int_z^{\infty} t^{-n} \cos t \, dt,$$

(3-12) $$S_n = y^{n-1} \int_z^{\infty} t^{-n} \sin t \, dt,$$

$$z = x_A y.$$

For $n = 1$, 2 and 3, expressions for C_n and S_n are presented in Table 2, for other n they can be easily calculated (see e.g. [47]). In the table the

Table 2.

n	C_n	S_n
1	$-\mathrm{ci}(z)$	$-\mathrm{si}(z)$
2	$y[-\mathrm{si}(z) + z^{-1} \cos z]$	$y[\mathrm{ci}(z) + z^{-1} \sin z]$
3	$y^2[\frac{1}{2}\mathrm{ci}(z) - (2z^2)^{-1} \cos z - (2z)^{-1} \sin z]$	$y^2[\frac{1}{2}\mathrm{si}(z) + (2z^2)^{-1} \sin z + (2z)^{-1} \cos z]$

following notation is used

$$\text{(3-13)} \qquad \mathrm{ci}(z) = -\int_z^{\infty} t^{-1} \cos t \,\mathrm{d}t,$$

$$\text{(3-14)} \qquad \mathrm{si}(z) = -\int_z^{\infty} t^{-1} \sin t \,\mathrm{d}t = -\frac{\pi}{2} + \mathrm{Si}(z).$$

3.1.5 Computation of the Fourier transform of a finite part of a function

Computing the Fourier transform of finite functions is the concern of the majority of numerical methods developed for computing Fourier transforms. A number of such methods, particularly older ones, is presented in [76]. More recent methods are included in [34], [57] and [58].

All these methods can be divided from a user's point of view into two groups, depending on whether a function f must be given in equidistant or non-equidistant points called nodes. Most recently recorded methods have used an equidistant step, even though from a practical viewpoint they are not necessarily the most suitable. Their use has extended by the high efficiency of an algorithm known as the fast Fourier transform (FFT).

Methods using a non-equidistant step possess the great advantage that the steplength can be adapted to the rate of change of a function f. It happens sometimes that a function f has considerably varying values over a range of several decades of a variable. An equidistant step method then requires a large number of values to express the function adequately, while function values in some regions change only very little. Non-equidistant methods allow one to describe the function f by a substantially smaller number of values. Nevertheless, methods using an equidistant step are often preferred, because function values may be obtained from automatic devices with a fixed step. Section 5.4 will be devoted to methods with an equidistant step using the discrete Fourier transform. Then, in the following section, we mention the non-equidistant methods.

There is a somewhat independent group of computing methods for which one must know the values of a function f at prescribed points. Examples of this are Salzer's method [57] and other methods based upon Gaussian integration. The applicability of these methods is, however, very limited and they are best suited to computing the Fourier transforms of functions which can be explicitly mathematically expressed in any point.

While a Fourier transform computation represents a numerical calculation of an integral with an oscillating integrand for any values of y, there are *a priori* advantages in methods interpolating $f(x)$ only and leaving the kernel e^{-jyx} unchanged. The method of interpolation gives us another way to classify numerical methods. Since only the values of the function at the nodes are known, the most suitable for practical use are methods in which just an interpolation from the known function values $f(x_i)$ is carried out. Also well known, however, are methods (see e.g. [57]), where an interpolation is carried out not only from values of a function but also from values of a derivative.

Another way to classify the interpolation methods is given by the character of the interpolating function. The most common interpolating functions are polynomials. This kind of interpolation has a lot of virtues resulting from a conversion to a spline interpolation. Some basic facts will be presented in § 3.2.3. The literature describes, however, methods using an exponential interpolation [57], [58] with an equidistant step and methods using non-equidistant interpolation by rational functions $(1 + x)^{-i}$ [57] applicable to a whole interval $(0, \infty)$. In principle, it is also possible to approach the numerical computation of Fourier transforms with some special methods for computing quickly oscillating integrals [34].

3.2 Computation of Fourier transforms by means of polynomial interpolation

3.2.1 The principle of polynomial interpolation

To be more concrete we shall further deal with the computation of Fourier transforms in the form:

$$F(\omega) = \int_{t_0}^{t_N} f(t)\, e^{-j\omega t}\, dt, \qquad -\infty < \omega < \infty; \tag{3-15}$$

at the same time we shall assume that $f(t)$ is expressed in terms of a function $\varphi_i(t)$ in the following way:

$$f(t) = \varphi_i(t), \qquad t \in (t_i, t_{i+1}), \quad i = 0, 1, 2, \ldots, N-1. \tag{3-16}$$

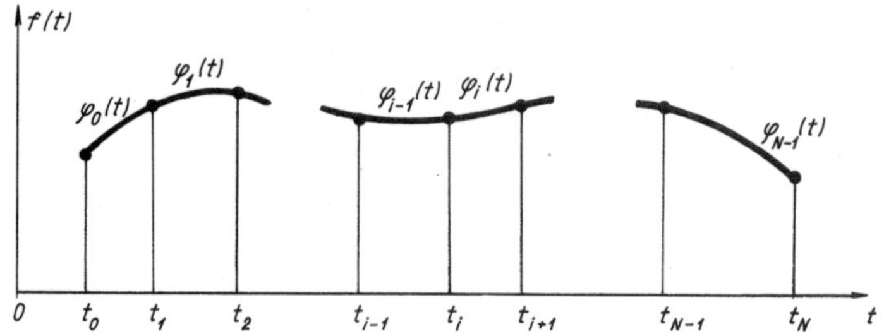

3. Relationship between a function f and functions φ_i

Concerning the functions φ_i, we shall assume that they are polynomials of n-th degree; hence they are continuous including their derivatives. By continuity at the terminal points of an interval (t_i, t_{i+1}) we shall understand continuity on the right at a point t_i and continuity on the left at point t_{i+1}. A graph of a function f with those properties is presented in Figure 3.

A further procedure is based upon the application of the following relation which can be obtained by an n-tuple integration by parts:

(3-17) $$\int f(t)\,\mathrm{e}^{-\mathrm{j}\omega t}\,\mathrm{d}t = \\ = \frac{1}{(\mathrm{j}\omega)^n}\int f^{(n)}(t)\,\mathrm{e}^{-\mathrm{j}\omega t}\,\mathrm{d}t - \sum_{\nu=0}^{n-1}\frac{f^{(\nu)}(t)}{(\mathrm{j}\omega)^{\nu+1}}\mathrm{e}^{-\mathrm{j}\omega t}, \qquad \omega \neq 0.$$

For the integral (3-15) where $f(t)$ is expressed in terms of $\varphi_i(t)$ we get for $N > 1$

(3-18) $$F(\omega) = \sum_{i=0}^{N-1}\left\{\frac{1}{(\mathrm{j}\omega)^n}\int_{t_i}^{t_{i+1}}\varphi_i^{(n)}(t)\,\mathrm{e}^{-\mathrm{j}\omega t}\,\mathrm{d}t - \right. \\ \left. - \sum_{\nu=0}^{n-1}\frac{1}{(\mathrm{j}\omega)^{\nu+1}}\left[\varphi_i^{(\nu)}(t_{i+1})\,\mathrm{e}^{-\mathrm{j}\omega t_{i+1}} - \varphi_i^{(\nu)}(t_i)\,\mathrm{e}^{-\mathrm{j}\omega t_i}\right]\right\}.$$

If we introduce a notation for differences of function values and derivatives at points t_i

(3-19) $$\varphi_i^{(\nu)}(t_i) - \varphi_{i-1}^{(\nu)}(t_i) = \Delta_i^{(\nu)}, \qquad i = 1, 2, \ldots, N-1, \\ \nu = 0, 1, \ldots, N-1,$$

we can adjust relation (3-18) as follows:

(3-20) $$F(\omega) = \frac{1}{(\mathrm{j}\omega)^n}\sum_{i=0}^{N-1}\int_{t_i}^{t_{i+1}}\varphi^{(n)}(t)\,\mathrm{e}^{-\mathrm{j}\omega t}\,\mathrm{d}t + \\ + \sum_{\nu=0}^{n-1}\frac{1}{(\mathrm{j}\omega)^{\nu+1}}\sum_{i=1}^{N-1}\left[\Delta_i^{(\nu)}\,\mathrm{e}^{-\mathrm{j}\omega t_i} + \right. \\ \left. + \varphi_0^{(\nu)}(t_0)\,\mathrm{e}^{-\mathrm{j}\omega t_0} - \varphi_{N-1}^{(\nu)}(t_N)\,\mathrm{e}^{-\mathrm{j}\omega t_N}\right].$$

Let us assume, furthermore, that $\varphi_i^{(n)}(t) = \text{constant}$. Then for the integral in (3-20),

(3-21) $$\sum_{i=0}^{N-1}\int_{t_i}^{t_{i+1}}\varphi_i^{(n)}(t)\,\mathrm{e}^{-\mathrm{j}\omega t}\,\mathrm{d}t = \\ = \frac{1}{\mathrm{j}\omega}\left[\sum_{i=1}^{N-1}\Delta_i^{(n)}\,\mathrm{e}^{-\mathrm{j}\omega t_i} + \varphi_0^{(n)}(t_0)\,\mathrm{e}^{-\mathrm{j}\omega t_0} - \varphi_{N-1}^{(n)}(t_N)\,\mathrm{e}^{-\mathrm{j}\omega t_N}\right]$$

and on this basis $F(\omega)$ can be expressed as

$$(3\text{-}22)\qquad F(\omega) = \frac{1}{(j\omega)^{n+1}} \sum_{i=1}^{N-1} \Delta_i^{(n)} e^{-j\omega t_i} + \sum_{v=0}^{n-1} \frac{1}{(j\omega)^{v+1}} \sum_{i=1}^{N-1} \Delta_i^{(v)} e^{-j\omega t_i} + \sum_{v=0}^{n} \frac{1}{(j\omega)^{v+1}} \left[\varphi_0^{(v)}(t_0) e^{-j\omega t_0} - \varphi_{N-1}^{(v)}(t_N) e^{-j\omega t_N}\right].$$

From this it follows that the Fourier transform of every function with the given properties, i.e. a function piecewise smooth up to order $n - 1$ and with derivative of n-th order piecewise constant, may be expressed at points t_i on the basis of steps of this function $\varphi_i(t) = \varphi^{(0)}(t)$ and steps of derivatives $\varphi_i^{(v)}(t)$, $v = 1, 2, \ldots, n$. As was already stated, the functions $\varphi_i(t)$ are n-th degree polynomials. This kind of expression was first described by A. Eagle [36] in 1928 and his method is used by a number of other authors [48], [28], [1].

If we formally introduce functions φ_{-1} and φ_N where we define $\varphi_{-1}(t) = 0$ and $\varphi_N(t) = 0$, then quantities

$$(3\text{-}23)\qquad \Delta_0^{(v)} = \varphi_0^{(v)}(t_0),$$

$$(3\text{-}24)\qquad \Delta_N^{(v)} = -\varphi_{N-1}^{(v)}(t_N)$$

can be added and relation (22) can be written in a compact form

$$(3\text{-}25)\qquad F(\omega) = \sum_{v=0}^{n} \frac{1}{(j\omega)^{v+1}} \sum_{i=0}^{N} \Delta_i^{(v)} e^{-j\omega t_i}.$$

This way of expressing the function f affords many possibilities for its practical application, some of which will be presented in the next section. None of these relations may be used for $\omega = 0$. For $\omega = 0$ the following relation

$$(3\text{-}26)\qquad F(0) = \int_{t_1}^{t_N} f(t)\, dt = \sum_{i=0}^{N-1} \int_{t_i}^{t_{i+1}} \varphi_i(t)\, dt$$

must be used or a function F at the point $\omega = 0$ must be defined by the limit

$$(3\text{-}27)\qquad F(0) = \lim_{\omega \to 0} F(\omega) = \sum_{v=0}^{n} (-1)^{v+1} \frac{1}{(v+1)!} \sum_{i=0}^{N} \Delta_i^{(v)} t_i^{v+1}.$$

The derived relations can also be used for an equidistant interpolation by means of functions φ_i. If we introduce

$$(3\text{-}28)\qquad t_i = iT, \qquad i = 0, 1, \ldots, N,$$

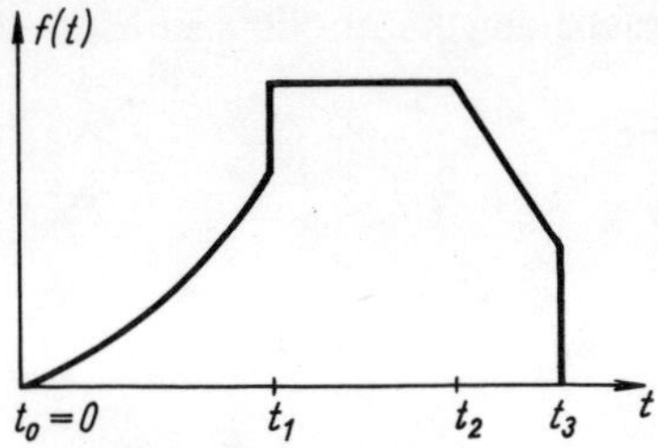

4a. Curve of the function f from example 1

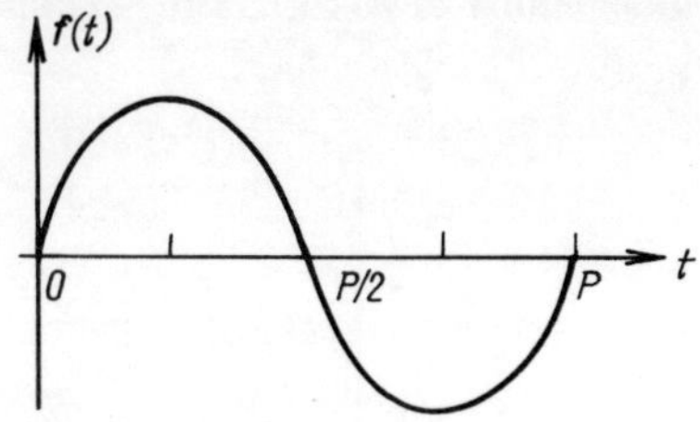

4b. Curve of the function f from example 2

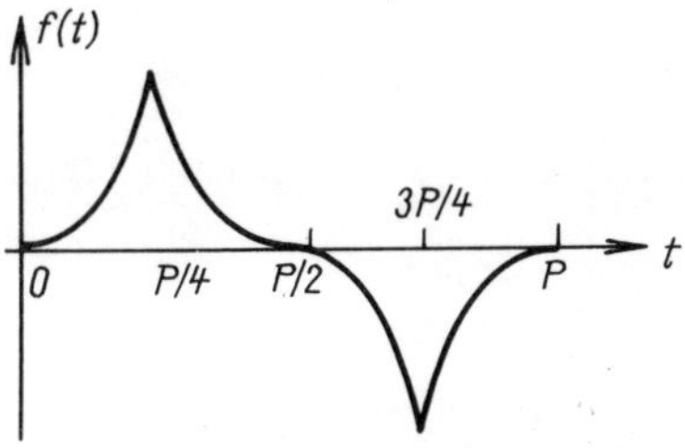

4c. Curve of the function f from example 3

then relations (3-25) and (3-27) become

$$(3\text{-}29)\qquad F(\omega) = \sum_{\nu=0}^{n} \frac{1}{(\mathrm{j}\omega)^{\nu+1}} \sum_{i=0}^{N} \Delta_i^{(\nu)} \mathrm{e}^{-\mathrm{j}\omega iT},$$

$$F(0) = \sum_{\nu=0}^{n} (-1)^{\nu+1} \frac{T^{\nu+1}}{(\nu+1)!} \sum_{i=0}^{N} \Delta_i^{(\nu)} i^{\nu+1}.$$

Example 1. Application of *Eagle's method.* A function f shown in Figure 4(a) is defined by

$$f(t) = \begin{cases} t^2, & t_0 \leqq t < t_1, \\ a, & t_1 < t < t_2, \\ kt + q, & t_2 \leqq t < t_3, \\ 0, & t < 0, \quad t > t_3, \end{cases}$$

$$k \neq 0, \quad q = a - kt_2, \quad t_0 = 0, \; t_3 = P.$$

For the values of jumps of the function f and their derivatives we get

$$\Delta_1 = a - t_1^2, \qquad \Delta_3 = -(kt_3 + q),$$
$$\Delta_1^{(1)} = -2t_i, \qquad \Delta_2^{(1)} = k, \quad \Delta_3^{(1)} = -k,$$
$$\Delta_0^{(2)} = 2, \qquad \Delta_1^{(2)} = -2.$$

Then for $F(\omega)$ using (3-22) we get

$$F(\omega) = -\mathrm{j}\omega^{-1}(\Delta_1 \mathrm{e}^{-\mathrm{j}\omega t_1} + \Delta_3 \mathrm{e}^{-\mathrm{j}\omega P}) -$$
$$- \omega^{-2}(\Delta_1^{(1)} \mathrm{e}^{-\mathrm{j}\omega t_1} + \Delta_2^{(1)} \mathrm{e}^{-\mathrm{j}\omega t_2} + \Delta_3^{(1)} \mathrm{e}^{-\mathrm{j}\omega P}) +$$
$$+ \mathrm{j}\omega^{-3}\Delta_0^{(2)}(1 - e^{-\mathrm{j}\omega P}).$$

Example 2. A function f composed from parabolic segments, drawn in Figure 4(b)

$$f(t) = \begin{cases} -16\dfrac{(t-P/4)^2}{P^2} + 1, & 0 \leqq t < \dfrac{P}{2}, \\ 16\dfrac{(t-3P/4)^2}{P^2} - 1, & \dfrac{P}{2} < t < P, \\ 0, & t < 0, \quad t > P, \end{cases}$$

has discontinuities just in the second order derivative at points $t_0 = 0$, $t_i = P/2$, $t_2 = P$ with these values of jumps:

$$\Delta_0^{(2)} = -32/P^2, \quad \Delta_1^{(2)} = 64/P^2, \quad \Delta_2^{(2)} = -32/P^2.$$

For the Fourier transform $F(\omega)$ we therefore obtain from relation (3-22)

$$F(\omega) = \mathrm{j}\frac{1}{\omega^3}(-1 + 2\mathrm{e}^{-\mathrm{j}\omega P/2} - \mathrm{e}^{-\mathrm{j}\omega P})\frac{32}{P^2}.$$

The coefficients C_k of the Fourier series, which arises by a periodic repetition of $f(t)$ with period P, are then determined according to (2-44), by the expression

$$C_k = \begin{cases} -\mathrm{j}\dfrac{16}{\pi^3 k^3}, & k \text{ odd}, \\ 0, & k \text{ even}, k = 0. \end{cases}$$

Example 3. The function f shown in Figure 4(c)

$$f(t) = \begin{cases} f_i(t), & 0 \leqq t < P/2, \\ -f_1(t - P/2), & P/2 < t < P, \\ 0, & t < 0, \quad t > P, \end{cases}$$

$$f_1(t) = \begin{cases} at^2, & 0 \leqq t < P/4, \\ a(t - P/2)^2, & P/4 < t < P/2, \end{cases}$$

has discontinuous derivatives at points $t_0 = 0$, $t_1 = P/4$, $t_2 = P/2$, $t_3 = 3P/4$, $t_4 = P$ with these jumps

$$\Delta_1^{(1)} = -2c, \quad \Delta_3^{(1)} = 2c,$$
$$\Delta_0^{(2)} = b, \quad \Delta_2^{(2)} = -2b, \quad \Delta_4^{(2)} = b,$$

where $b = 2a$, $c = aP/2$, $a \neq 0$.

For the Fourier transform $F(\omega)$ we get

$$F(\omega) = -\frac{2c}{\omega^2}[-\mathrm{e}^{-\mathrm{j}\omega P/4} + \mathrm{e}^{-\mathrm{j}\omega 3P/4}] +$$
$$+ \mathrm{j}\frac{b}{\omega^3}[1 - 2\mathrm{e}^{-\mathrm{j}\omega P/2} + \mathrm{e}^{-\mathrm{j}\omega P}]$$

and the following holds for the coefficients C_k of the Fourier series of the periodised function f

$$C_k = \begin{cases} 0, & k \text{ even}, \\ \dfrac{4}{P}\left(\mathrm{j}\dfrac{b}{\omega_k^3} + \dfrac{c}{\omega_k^2}\,\mathrm{e}^{-\mathrm{j}k\pi/2}\right), & k \text{ odd}, \end{cases}$$

$$\omega_k = \frac{k\,.\,2\pi}{P}.$$

3.2.2 Use of a polynomial interpolation

We now demonstrate how the derived relation can be used in the case where φ_i are polynomials of different degree and different properties in interpolating nodes.

A. Interpolation by polynomials

In this case we choose for φ_i polynomials of n-th degree going through points $t_i, f_i, t_{i+1}, f_{i+1}, t_{i+2}, f_{i+2}, \ldots, t_{i+\mu}, f_{i+\mu}, \mu = 1, 2, \ldots, n$. At the same time we choose N so that N/n is an integer, so as to divide an interval $(t_0; t_N)$ into an integral number of partial intervals.

(a) $n = 0$: Interpolation by a step function

This trivial kind of interpolation corresponds to a step value of a function φ_i between single values f_i, while $\varphi_i(t) = f_i$ for all $t \in [t_i, t_{i+1})$; a graph of this is shown in Figure 5.

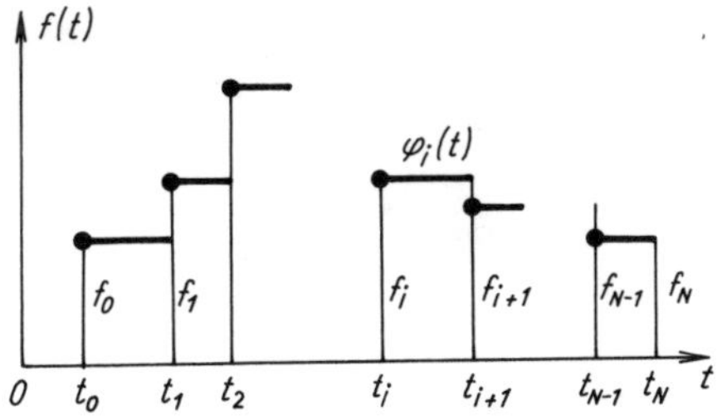

5. Stepwise interpolation of a function f

From relations (3-25) and (3-26) we get for $n = 0$ and $\Delta_i^{(0)} = \Delta_i$

$$\text{(3-30)} \qquad F(\omega) = \frac{1}{\mathrm{j}\omega} \sum_{i=0}^{N} \Delta_i \,\mathrm{e}^{-\mathrm{j}\omega t_i}, \qquad \omega \neq 0,$$

$$\text{(3-31)} \qquad F(0) = \sum_{i=0}^{N-1} f_i(t_{i+1} - t_i),$$

where $\Delta_i = f_i - f_{i-1}$, $\Delta_0 = f_0$, $\Delta_N = -f_{N-1}$.

For the equidistant case we get

$$F(\omega) = \frac{1}{\mathrm{j}\omega} \sum_{i=0}^{N} \Delta_i \, \mathrm{e}^{-\mathrm{j}\omega i T}, \qquad \omega \neq 0, \tag{3-32}$$

$$F(0) = T \sum_{i=0}^{N-1} f_i . \tag{3-32a}$$

If we substitute for Δ_i, the above relations can be written as:

$$F(\omega) = T\mathrm{e}^{-\mathrm{j}\omega T/2} \varrho\left(\frac{\omega T}{2}\right) \sum_{i=0}^{N-1} f_i \, \mathrm{e}^{-\mathrm{j}\omega i T}, \tag{3-33}$$

where

$$\varrho(x) = x^{-1} \sin x. \tag{3-34}$$

An interpolation by a step function thus appears in a Fourier transform as a weight function $\varrho(\omega T/2)$.

(b) $n = 1$: Interpolation by a piecewise linear function

In this case we carry out an interpolation of a function f by a set of segments of straight lines which create a broken line as shown in Figure 6. For φ_i there holds:

$$\varphi_i(t) = k_i(t - t_i) + f_i,$$

where

$$k_i = \frac{f_{i+1} - f_i}{t_{i+1} - t_i}, \qquad i = 1, 2, \ldots, N - 1.$$

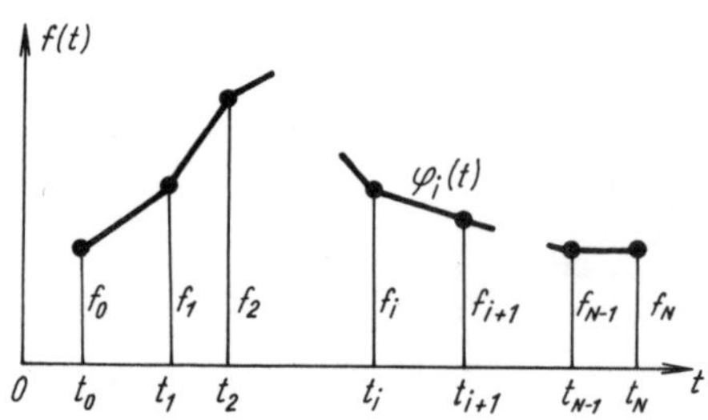

6. Piecewise linear interpolation of a function f

For $n = 1$ we obtain from (3-22)

$$F(\omega) = \frac{1}{(\mathrm{j}\omega)^2} \times \tag{3-35}$$

$$\times \left[\sum_{i=1}^{N-1} \Delta_i^{(1)} \mathrm{e}^{-\mathrm{j}\omega t_i} + \varphi_0^{(1)}(t_0) \, \mathrm{e}^{-\mathrm{j}\omega t_0} - \varphi_{N-1}^{(1)}(t_N) \, \mathrm{e}^{-\mathrm{j}\omega t_N} \right] +$$

$$+ \frac{1}{\mathrm{j}\omega} \left[f_0 \, \mathrm{e}^{-\mathrm{j}\omega t_0} - f_N \, \mathrm{e}^{-\mathrm{j}\omega t_N} \right].$$

If we replace

$$\Delta_i^{(1)} \quad = k_i - k_{i-1},$$

$$\varphi_0^{(1)}(t_0) \quad = k_0,$$

$$\varphi_{N-1}^{(1)}(t_N) = k_{N-1},$$

then after some manipulation we arrive at the expression

$$\text{(3-36)} \quad F(\omega) = \frac{1}{\omega^2} \sum_{i=0}^{N-1} k_i (\mathrm{e}^{-\mathrm{j}\omega t_{i+1}} - \mathrm{e}^{-\mathrm{j}\omega t_i}) + \\ + \mathrm{j}\frac{1}{\omega}(f_N \mathrm{e}^{-\mathrm{j}\omega t_N} - f_0 \mathrm{e}^{-\mathrm{j}\omega t_0}), \qquad \omega \neq 0,$$

$$\text{(3-37)} \quad F(0) = \frac{1}{2} \sum_{i=0}^{N-1} (f_{i+1} + f_i)(t_{i+1} - t_i).$$

Using an equidistant interpolation $t_i = iT$, $i = 0, 1, \ldots, N$, we get

$$\text{(3-38)} \quad F(\omega) = \frac{1}{\omega^2} \sum_{i=0}^{N-1} k_i \mathrm{e}^{-\mathrm{j}\omega iT} (\mathrm{e}^{-\mathrm{j}\omega T} - 1) + \\ + \mathrm{j}\frac{1}{\omega}(f_N \mathrm{e}^{-\mathrm{j}\omega NT} - f_0),$$

$$\text{(3-39)} \quad F(0) = \frac{T}{2} \sum_{i=0}^{N-1} (f_{i+1} + f_i).$$

If we substitute $k_i = T^{-1}(f_{i+1} - f_i)$, then (3-38) may be adjusted to this form:

$$\text{(3-40)} \quad F(\omega) = T\varrho^2\left(\frac{1}{2}\omega T\right) \sum_{i=0}^{N-1} f_i \mathrm{e}^{-\mathrm{j}\omega iT} + \\ + \left(\frac{1}{T}\frac{1 - \mathrm{e}^{-\mathrm{j}\omega T}}{\omega^2} + \mathrm{j}\frac{1}{\omega}\right)(f_N \mathrm{e}^{-\mathrm{j}\omega NT} - f_0),$$

where ϱ is the weight function determined by (3-34).

(c) Generally, it would be possible to use interpolation by higher-degree polynomials. An interpolation polynomial of n-th degree requires for its specification at least $n + 1$ function values, which is why in this case N must be an integer multiple of n and functions φ_i are specified on intervals $[t_i, t_{i+n})$. A drawback of this method of interpolation is that it does not guarantee continuity at points of transition between particular definition intervals for functions φ_i. To this group of interpolation methods belongs the well known Filon's method [40], corresponding to $n = 2$. Relations for these methods are given for $n = 2, 3, 4$ in [57], [58].

3.2.3 Application of a spline interpolation

Relation (3-25) containing an expression for the difference of derivatives at nodes t_i is very suitable for a spline interpolation [31], [36]. By the *spline polynomial interpolation* of n-th degree [3], [34] we understand an interpolation with nodes t_i, f_i, $i = 0, 1, \ldots, N$, by a function that is an n-th degree polynomial φ_i in each interval $[t_i, t_{i+1}]$, $i = 0, 1, \ldots, N - 1$. Polynomials φ_i, $i = 0, 1, \ldots, N - 1$ satisfy an interpolation condition

$$\varphi_i(t_i) = \varphi_{i-1}(t_i) = f_i, \qquad i = 1, 2, \ldots, N - 1,$$

and, furthermore, have in their inner nodes continuous derivatives up to order $(n - 1)$, i.e. they satisfy

$$\varphi_i^{(v)}(t_i) = \varphi_{i-1}^{(v)}(t_i v), \qquad v = 1, 2, \ldots, n - 1, \quad i = 1, 2, \ldots, N - 1.$$

To these conditions should be added conditions for the behaviour of the interpolation polynomials at terminal nodes t_0, f_0 and t_N, f_N.

Considering the presented conditions, then for our case must hold

$$\text{(3-41)} \qquad \Delta_i^{(v)} = 0, \qquad v = 0, 1, \ldots, n - 1, \quad i = 1, 2, \ldots, N - 1,$$

and relation (3-22) converts to the form

$$\text{(3-42a)} \quad F(\omega) = \frac{1}{(j\omega)^{n+1}} \sum_{i=1}^{N-1} \Delta_i^{(n)} e^{-j\omega t_i} + \sum_{v=0}^{n} \frac{1}{(j\omega)^{v+1}} [\varphi_0^{(v)}(t_0) e^{-j\omega t_0} - \varphi_{N-1}^{(v)}(t_N) e^{-j\omega t_N}]$$

or to the equivalent form

$$\text{(3-42b)} \quad F(\omega) = \frac{1}{(j\omega)^{v+1}} \sum_{i=0}^{N} \Delta_i^{(n)} e^{-j\omega t_i} + \sum_{v=0}^{n-1} \frac{1}{(j\omega)^{v+1}} [\Delta_0^{(v)} e^{-j\omega t_0} + \Delta_N^{(v)} e^{-j\omega t_N}],$$

where

$$\Delta_0^v = \varphi_0^{(v)}(t_0),$$
$$\Delta_N^v = -\varphi_{N-1}^{(v)}(t_N),$$
$$\Delta_i^v = \varphi_i^{(v)}(t_i) - \varphi_{i-1}^{(v)}(t_i), \qquad i = 1, 2, \ldots, N - 1, \quad v = 0, 1, \ldots, n.$$

On using the most usual spline interpolation of third degree, i.e. $n = 3$, one gets

$$\text{(3-43)} \quad F(\omega) = \frac{1}{(j\omega)^4} \sum_{i=1}^{N-1} \Delta_i^{(3)} e^{-j\omega t_i} + \sum_{v=0}^{3} \frac{1}{(j\omega)^{v+1}} [\varphi_0^{(v)}(t_0) e^{-j\omega t_0} - \varphi_{N-1}^{(v)}(t_N) e^{-j\omega t_N}].$$

Using relations for spline interpolation from [3], we obtain

$$\begin{aligned}
\Delta_i^{(3)} &= M_{i+1} - 2M_i + M_{i-1}, \qquad i = 1, 2, \ldots, N-1,\\
\varphi_0(t_0) &= f_0,\\
\varphi_{N-1}(t_N) &= f_N,\\
\varphi_0^{(1)}(t_0) &= -\frac{h_1}{3} M_0 - \frac{h_1}{6} M_1 + \frac{f_1 - f_0}{h_1},\\
\varphi_{N-1}^{(1)}(t_N) &= \frac{h_N}{6} M_{N-1} + \frac{h_N}{3} M_N + \frac{f_N - f_{N-1}}{f_N}\\
\varphi_0^{(2)}(t_0) &= M_0,\\
\varphi_{N-1}^{(2)}(t_N) &= M_N,\\
\varphi_0^{(3)}(t_0) &= M_1 - M_0,\\
\varphi_{N-1}^{(3)}(t_N) &= M_N - M_{N-1}.
\end{aligned} \tag{3-44}$$

The quantities M_i (moments) of a spline interpolation are determined by the solution of a tridiagonal set of equations

$$\left.\begin{aligned}
2M_0 + \lambda_0 M_1 &= d_0,\\
\mu_i M_{i-1} + 2M_i + \lambda_i M_{i+1} &= d_i,\\
\lambda_N M_{N-1} + 2M_N &= d_N,
\end{aligned}\right\} \quad i = 1, 2, \ldots, N-1, \tag{3-45}$$

with notation

$$\begin{aligned}
&\left.\begin{aligned}
h_i &= t_i - t_{i-1},\\
\sigma_i &= h_i^{-1}(f_i - f_{i-1}),
\end{aligned}\right\} \quad i = 1, 2, \ldots, N.\\
&\left.\begin{aligned}
\lambda_i &= h_{i+1}(h_i + h_{i+1})^{-1},\\
\mu_i &= 1 - \lambda_i,\\
d_i &= 6(h_i + h_{i+1})^{-1}(\sigma_{i+1} - \sigma_i).
\end{aligned}\right\} \quad i = 1, 2, \ldots, N-1,
\end{aligned} \tag{3-46}$$

For $\omega = 0$ the following relation can be derived:

$$F(0) = \sum_{i=1}^{N} \left[\frac{1}{2} h_i(f_i + f_{i-1}) - \frac{1}{24} h_i^3(M_i + M_{i-1}) \right]. \tag{3-47}$$

The quantities λ_0, d_0, λ_N, d_N, M_0 and M_N should be given specifically depending on the choice of type of spline interpolation.

a) Natural spline

When using so-called natural splines, $d_0 = \lambda_0 = d_N = \lambda_N = 0$ and therefore $M_0 = M_N = 0$. The remaining M_i, $i = 1, 2, \ldots, N-1$ are determined from the system of equations expressed in relation (3-45) in a recurrent manner. As a result the second derivative is equal to zero at the boundary points

t_0 and t_N, i.e. $\varphi^{(2)}(t_0) = 0$ and $\varphi^{(2)}_{N-1}(t_N) = 0$. For a Fourier transform we get from relation (3-43):

$$(3\text{-}48)\qquad F(\omega) = \frac{1}{(\mathrm{j}\omega)^4}\Big[\sum_{i=1}^{N-1} \Delta_i^{(3)}\,\mathrm{e}^{-\mathrm{j}\omega t_i} + \varphi_0^{(3)}(t_0)\,\mathrm{e}^{-\mathrm{j}\omega t_0} - \varphi_{N-1}^{(3)}(t_N)\,\mathrm{e}^{-\mathrm{j}\omega t_N}\Big] + \frac{1}{(\mathrm{j}\omega)^2}\Big[\varphi_0^{(1)}(t_0)\,\mathrm{e}^{-\mathrm{j}\omega t_0} - \varphi_{N-1}^{(1)}(t_N)\,\mathrm{e}^{-\mathrm{j}\omega t_N}\Big] + \frac{1}{\mathrm{j}\omega}\Big[f_0\,\mathrm{e}^{-\mathrm{j}\omega t_0} - f_N\,\mathrm{e}^{-\mathrm{j}\omega t_N}\Big].$$

b) Periodic spline

We shall consider a periodic spline satisfying the boundary conditions

$$(3\text{-}49)\qquad f_0 = f_N,$$

$$(3\text{-}50)\qquad \varphi_0^{(\nu)}(t_0) = \varphi_{N-1}^{(\nu)}(t_N), \qquad \nu = 1, 2.$$

Then for the Fourier transform we get from (3-43)

$$(3\text{-}51)\qquad F(\omega) = \frac{1}{(\mathrm{j}\omega)^4}\sum_{i=1}^{N-1} \Delta_i^{(3)}\,\mathrm{e}^{-\mathrm{j}\omega t_i} + \left[\frac{1}{\mathrm{j}\omega} f_0 + \frac{1}{(\mathrm{j}\omega)^2}\varphi_0^{(1)}(t_0) + \frac{1}{(\mathrm{j}\omega)^3}\varphi_0^{(2)}(t_0)\right] \times \left[\mathrm{e}^{-\mathrm{j}\omega t_0} - \mathrm{e}^{-\mathrm{j}\omega t_N}\right] + \frac{1}{(\mathrm{j}\omega)^4}\left[\varphi_0^{(3)}(t_0)\,\mathrm{e}^{-\mathrm{j}\omega t_0} - \varphi_{N-1}^{(3)}(t_0)\,\mathrm{e}^{-\mathrm{j}\omega t_N}\right],$$

while (3.44) holds.

Quantities M_i, $i = 0, 1, \ldots, N$ are determined by the solution of a set of equations

$$(3\text{-}52)\qquad \begin{aligned} &\mu_i M_{i-1} + 2M_i + \lambda_i M_{i+1} = d_i, \qquad i = 1, 2, \ldots, N-1,\\ &\lambda_N M_1 + \mu_N M_{N-1} + 2M_N = d_N,\\ &M_N = M_0. \end{aligned}$$

For h_i, σ_i, λ_i and d_i relations (3-46) and relations

$$(3\text{-}53)\qquad \begin{aligned} &\lambda_N = \frac{h_1}{h_1 + h_N},\\ &\mu_N = 1 - \lambda_N,\\ &\mathrm{d}_N = \frac{6}{h_1 + h_N}\left[\frac{f_1 - f_N}{h_1} + \frac{f_N - f_{N-1}}{h_N}\right] \end{aligned}$$

hold.

3.3 Elimination of the effect of function discontinuities

3.3.1 Effect of function discontinuities and their derivatives on the computation of Fourier transforms

We have stated in § 2.1.3 how the rate of decrease of absolute values of coefficients of the Fourier series depends with growing k on the smoothness of a function and in § 2.4.1 we have shown the same for the Fourier transform. We have also illustrated by Eagle's method that a discontinuity of the ν-th order derivative of a function x, $\nu = 0, 1, \ldots$, is shown by the presence of terms $(j\omega)^{-\nu-1}$ in its Fourier transform. For the sake of satisfying numerical requirements in computing the Fourier transform, it is advantageous that its absolute value for $|\omega| \to \infty$ decreases rapidly. For this reason, it is useful in numerical computation of the Fourier transform of a function with discontinuities to separate out from a given function well chosen auxiliary functions in which all discontinuities will be included. Numerical computation is then carried out with a function that is already quite smooth. Auxiliary functions should be chosen so that their Fourier transforms can be determined by direct computation, i.e. without numerical methods. Such a procedure will be described in § 3.3.2.

If a function f is known only as a sequence of numerical values, only the discontinuities of a function and of its first order derivatives can usually be clearly distinguished. If we eliminate these discontinuities, the Fourier transform of a smoothed out function will contain as its most slowly decreasing term $(j\omega)^{-3}$ and this will be very advantageous for numerical computation.

3.3.2 Krylov's method of elimination of discontinuities

Let us begin with a function f having in a periodicity interval $(0, P)$ a discontinuity of an order ν at a point t_i. Krylov's method [57], [59] describes the construction of a function h which should be subtracted from f on the periodicity interval in order to eliminate the discontinuity. We thereby obtain a function g,

$$g(t) = f(t) - h(t), \qquad t \in [0, P), \tag{3-54}$$

continuous at the point t_i which will then be periodically extended. As the position of a point t_i can easily be modified by shifting, we further simply consider $t_i = 0$.

Let us consider now a periodic function f having at a point $t_i = 0$ a jump of size $\Delta_0 = f(0) - f(P)$. To eliminate this jump, we subtract from f a function $h_0 = \Delta_0 \gamma_0$, where

$$\gamma_0(t) = -\frac{1}{P} t + \frac{1}{2}, \qquad t \in [0, P). \tag{3-55}$$

Then, for the function g

$$g(0) = g(P) = \frac{1}{2}[f(P) + f(0)]. \tag{3-56}$$

If the function f has at a point $t_i = 0$ a jump in a derivative of size $\Delta_0^{(1)} = f'(0) - f'(P)$, to eliminate it we must subtract from f' a function $\Delta_0^{(1)}\gamma_0$, therefore from a function f a function $h_1 = \Delta_0^{(1)}\gamma_1$ where

$$\gamma_1(t) = \int_0^t \gamma_0(t)\,\mathrm{d}t + c_1. \tag{3-57}$$

Choosing an additional condition for determining the constant c_1, e.g. of the form

$$\int_0^P \gamma_1(t)\,\mathrm{d}t = 0, \tag{3-58}$$

one gets

$$\gamma_1(t) = -(1/2P)\,t^2 + \frac{1}{2}t - \frac{1}{12}P. \tag{3-59}$$

In a similar way, for the elimination of a jump in a second order derivative at a point $t_i = 0$ of size $\Delta_0^{(2)}$, we arrive at a function $h_2 = \Delta_0^{(2)}\gamma_2$, where

$$\gamma_2(t) = -(1/6P)\,t^3 + \frac{1}{4}t^2 - (P/12)\,t. \tag{3-60}$$

We can carry on in this way and so eliminate discontinuities of a derivative of any order at a point $t_i = 0$.

If a function f has a discontinuity of a ν-th order derivative at a general inner point $t_1 \in (0, P)$, then we must subtract a function $\Delta_i^{(\nu)}\gamma_{\nu,t_i}$, where

$$\gamma_{\nu,t_i}(t) = \begin{cases} \gamma_\nu(t - t_i), & t_i < t < P, \\ \gamma_\nu(t + P - t_i), & 0 < t < t_i. \end{cases} \tag{3-61}$$

If a function f has at each of the points t_i, $i = 0, 1, \ldots, n$ discontinuities in itself and in derivatives up to m_i-th order, we must subtract a function h where

$$h(t) = \sum_{i=0}^{n} \sum_{\nu=0}^{m_i} \Delta_i^{(\nu)}\gamma_{\nu,t_i}(t). \tag{3-62}$$

The Fourier transform of the function h is easily obtained by means of Eagle's method.

A function h constructed in this manner enables us to create periodic functions continuous up to any order at a point $t = 0$, i.e. functions for which $g^{(\nu)}(0) = g^{(\nu)}(P)$, $\nu = 0, 1, \ldots$. In the case of computation of a Fourier transform of a finite function which is nonzero on an interval $(0, P)$ this means, however, that a function g obtained in the way described, together with its derivatives, can possess discontinuities at points $t = 0$ and $t = P$.

If we want to eliminate these discontinuities, we first eliminate by Krylov's method all inner discontinuities. Thus we obtain a function g which has discontinuities at points $t = 0$ and $t = P$ only. Now, if we want to adjust g so that it has zero values together with its derivatives up to the m-th order at the mentioned points, we must subtract from it a Hermite interpolation polynomial [52] k subject to the conditions

$$k^{(\nu)}(0) = g^{(\nu)}(0) \quad \text{and} \quad k^{(\nu)}(P) = g^{(\nu)}(P) \tag{3-63}$$

for $\nu = 0, 1, \ldots, m$.

We thus create a finite function

$$p(t) = g(t) - k(t), \tag{3-64}$$

which is continuous together with its derivatives up to the m-th order at the points $t = 0$ and $t = P$.

If we require just a zero function value $p(t)$ for $t = 0$ and $t = P$, i.e. when $m = 0$, then for the function k holds

$$k(t) = \frac{g(P) - g(0)}{P} t + g(0), \qquad t \in [0, P). \tag{3-65}$$

If we want a function p to possess zero value of the function as well as zero value of the first derivative at points $t = 0$ and $t = P$, then for $k(t)$ the following relation can be derived:

$$\begin{aligned} k(t) = {} & \left[g(0) \left(1 + \frac{2t}{P} \right) + g'(0)\, t \right] \frac{(t - P)^2}{P^2} + \\ & + \left[g(P) \left(1 - 2 \frac{t - P}{P} \right) + g'(P)(t - P) \right] \frac{t^2}{P^2}, \\ & \hphantom{+} \qquad t \in [0, P]. \end{aligned} \tag{3-66}$$

We then compute the Fourier transform of the function k by means of Eagle's method.

Example 4. A function k expressed by relation (3-65) has jumps at points $t_0 = 0$ and $t_N = t_1 = P$ only:

$$\Delta_0 = g_0, \qquad \Delta_1 = -g_1,$$
$$\Delta_0^{(1)} = k', \qquad \Delta_1^{(1)} = -k',$$

where $g_0 = g(0)$, $g_1 = g(P)$ and $k' = (g_1 - g_0)/P$. By relation (3-22) there results for the Fourier transform $K(\omega)$ a relation

$$K(\omega) = -\mathrm{j} \frac{g_0 - g_1 \mathrm{e}^{-\mathrm{j}\omega P}}{\omega} - k' \frac{1 - \mathrm{e}^{-\mathrm{j}\omega P}}{\omega^2}.$$

Example 5. A function k expressed by relation (3-66) has jumps at points $t_0 = 0$ and $t_N = t_1 = P$:

$$\Delta_0 = g_0, \qquad \Delta_1 = -g_1,$$

$$\Delta_0^{(1)} = g'_0, \qquad \Delta_1^{(1)} = -g'_1,$$
$$\Delta_0^{(2)} = k''_0, \qquad \Delta_1^{(2)} = k''_1,$$
$$\Delta_0^{(3)} = k''', \qquad \Delta_1^{(3)} = -k'''.$$

Here we have used $g_0 = g(0)$, $g_1 = g(P)$, $g'_0 = g'(0)$, $g'_1 = g'(P)$,

$$k''_0 = 6\frac{g_1 - g_0}{P^2} - 2\frac{2g'_0 + g'_1}{P},$$

$$k''_1 = -6\frac{g_1 - g_0}{P^2} + 2\frac{2g'_1 + g'_0}{P},$$

$$k''' = 6\frac{2(g_0 - g_1)/P + g'_0 + g'_1}{P^2}.$$

For the Fourier transform of the function k expressed by relation (3-66) we then get

$$K(\omega) = -\mathrm{j}\frac{g_0 - g_1\,\mathrm{e}^{-\mathrm{j}\omega P}}{\omega} - \frac{g'_0 - g'_1\,\mathrm{e}^{-\mathrm{j}\omega P}}{\omega^2} + $$
$$+\,\mathrm{j}\frac{k''_0 - k''_1\,\mathrm{e}^{-\mathrm{j}\omega P}}{\omega^3} + k'''\frac{1 - \mathrm{e}^{-\mathrm{j}\omega P}}{\omega^4}.$$

4. DISCRETE FOURIER TRANSFORMS

Discrete Fourier transforms are transforms of finite sequences of complex or real numbers. The results are again finite sequences of generally complex numbers. Discrete Fourier transforms map a finite sequence of numbers onto a finite sequence of numbers. Under the term discrete Fourier transform we shall understand the direct discrete Fourier transform as well as the inverse discrete Fourier transform which maps the result of a direct transform back to the original sequence.

The Fourier transform is a relatively new transform. As a both-directions transform it was introduced first in 1967. However, relations defining a direct transform had already been used for a long time for the approximate computation of Fourier series coefficients and in a real transcription they are known as Bessel relations. A major development of discrete Fourier transforms appeared in 1965, when J. W. Cooley and J. W. Tukey [25] described a very fast algorithm for computation known as the fast Fourier transform (FFT). Thanks to this algorithm the discrete Fourier transform became the most widespread means of numerical computation of the Fourier transform and of numerical harmonic analysis.

4.1 Definition of discrete Fourier transforms and their properties

A finite sequence of elements z_i, $i = 0, 1, \ldots, N-1$, will be represented in the text by $z_0, z_1, \ldots, z_{N-1}$ or briefly just by z_i, if there is no possibility of misunderstanding. A symbol z_i thus will mean either the i-th element of a sequence or the whole sequence. For a discrete direct Fourier transform, we shall always represent the original sequence by small letters, its image by the same capital letter.

4.1.1 Definition of discrete Fourier transforms

Definition of direct discrete Fourier transforms

If x_i, $i = 0, 1, 2, \ldots, N-1$ is a sequence of N finite complex numbers, then its *direct discrete Fourier transform* or *a transform* $\mathscr{D}_N$ is given by a sequence of values X_k, where

$$X_k = \sum_{i=0}^{N-1} x_i \, \mathrm{e}^{-\mathrm{j}ik2\pi/N}, \qquad k = 0, 1, \ldots, N-1. \tag{4-1}$$

We shall call the sequence X_k *the image* or *transform* $\mathscr{D}_N$ of the sequence x_i.

Definition of inverse Fourier transforms

If X_k, $k = 0, 1, \ldots, N-1$ is a sequence of N finite complex numbers, then its *inverse discrete Fourier transform* or *a transform* $\mathscr{D}_N^{-1}$ is a sequence of N numbers expressed by the relation

$$x_i = \frac{1}{N} \sum_{k=0}^{N-1} X_k \, \mathrm{e}^{\mathrm{j}ik2\pi/N}, \qquad i = 0, 1, \ldots, N-1. \tag{4-2}$$

In the case where a sequence X_k is an image of a sequence x_i, the relation (4-2) expresses the original sequence x_i and is inverse to the relation (4-1). It follows by substituting (4-1) in (4-2),

$$x_i = \sum_{\nu=0}^{N-1} x_\nu \frac{1}{N} \sum_{k=0}^{N-1} \mathrm{e}^{\mathrm{j}k(i-\nu)2\pi/N},$$

since

$$\frac{1}{N} \sum_{k=0}^{N-1} \mathrm{e}^{\mathrm{j}k(i-\nu)2\pi/N} = \begin{cases} 0, & \nu \neq i (\mathrm{mod}\, N), \\ 1, & \nu = i (\mathrm{mod}\, N). \end{cases} \tag{4-3}$$

This relation expresses the discrete orthogonality of an exponential sequence $\mathrm{e}^{\mathrm{j}ik2\pi/N}$, $i = 0, 1, \ldots, N-1$.

We introduce also a symbolic representation of the direct transform

$$X_k = \mathscr{D}_n\{x_i\} \tag{4-4}$$

and of the inverse transform

$$x_i = \mathscr{D}_N^{-1}\{X_k\}. \tag{4-5}$$

In addition, we shall also use a simplified notation to express the mutual assignment of sequences x_i and X_k, in the form

$$x_i \mathrel{\hat{=}} X_k. \tag{4-6}$$

To simplify the notation further we introduce

$$\mathrm{e}^{-\mathrm{j}2\pi/N} = W_N. \tag{4-7}$$

So long as it is not necessary to emphasize the number of elements of a sequence we shall use symbols $\mathscr{D}$ instead of $\mathscr{D}_N$ and W instead of W_N, respectively.

The transform expressed by relations (4-1) and (4-2) is called *the discrete Fourier transform.* This name is not quite fitting; a more accurate name would be the finite discrete Fourier transform. However, we keep the original term, because it is used universally in English, French, German and Russian. Some authors also call this transform the finite Fourier transform (see e.g. [23]). In this chapter we shall just use the term transform, where no confusion is likely to arise.

The definition given for the discrete Fourier transform is used in almost all new publications. There do exist some other definitions, mostly older ones, differing only in a multiplicative coefficient or different sign of the exponent.

Both definition relations (4-1) and (4-2) express the unique assignment of one sequence to the other. That is, $\mathscr{D}_N$ and its inverse $\mathscr{D}_N^{-1}$ are well-defined operations.

Every definition sum, however, always assigns to a finite sequence an infinite periodic sequence (of period N) made up by repeating the basic period defined for $k = 0, 1, \ldots, N-1$, in the case of the $\mathscr{D}_N$ transform, or for $i = 0, 1, \ldots, N-1$, in the case of the $\mathscr{D}_N^{-1}$ transform. Since it is advantageous to use the periodic sequences, because of their close relationship with periodic functions, let us introduce the extended definition of the transforms $\mathscr{D}$ and $\mathscr{D}^{-1}$.

$$X_k = \sum_{i=0}^{N-1} x_i W^{ik}, \tag{4-8}$$

$$k = \varkappa + \nu N, \qquad \varkappa = 0, 1, \ldots, N-1, \qquad \nu = 0, \pm 1, \pm 2, \ldots,$$

$$x_i = \frac{1}{N} \sum_{k=0}^{N-1} X_k W^{-ik}, \tag{4-9}$$

$$i = \lambda + \mu N, \qquad \lambda = 0, 1, \ldots, N-1, \qquad \mu = 0, \pm 1, \pm 2, \ldots.$$

Hence, the discrete Fourier transforms can be considered as well-defined operations, either as assignments between finite sequences which define basic intervals of periodicity, or as assignments of infinite periodic sequences.

4.1.2 Basic properties of discrete Fourier transforms

In this chapter we present some basic properties of discrete Fourier transforms. We shall be concerned with quite elementary properties, and shall present them, almost without proof, where they are trivial and result mostly from the definitions and properties above. To simplify matters, we present theorems in the form of implications, although it is obvious that where they hold in one sense, they always hold in the other.

Theorem 1: Periodic properties of a transform and an inverse transform

The discrete Fourier transform and inverse discrete Fourier transform define periodic sequences

$$x_i = x_{i+\mu N}, \qquad \mu = 0, \pm 1, \pm 2, \ldots, \tag{4-10}$$

$$X_k = X_{k+\nu N}, \qquad \nu = 0, \pm 1, \pm 2, \ldots. \tag{4-11}$$

This assertion results directly from the defining relations.

As a special case, we get from relations (4-10) and (4-11) expressions for negative indices

$$x_{-i} = x_{N-i}, \tag{4-12}$$

$$X_{-k} = X_{N-k}. \tag{4-13}$$

Theorem 2: Symmetry of a transform

From the defining relations it follows that a direct transform can be expressed by means of an inverse and vice versa. If we mark complex conjugate values by an asterisk, we get

$$x_i = \left(\sum_{k=0}^{N-1} \frac{1}{N} X_k^* \, \mathrm{e}^{-\mathrm{j}ik2\pi/N} \right)^* = \left(\mathscr{D}\left\{ \frac{1}{N} X_k^* \right\} \right)^*, \tag{4-14}$$

$$X_k = \left(\frac{1}{N} \sum_{i=0}^{N-1} N x_i^* \, \mathrm{e}^{\mathrm{j}ik2\pi/N} \right)^* = (\mathscr{D}^{-1}\{N x_i^*\})^*. \tag{4-15}$$

The inverse transform may be expressed as the complex conjugate of the $(1/N)$ multiple of the direct transform applied to a complex conjugate image. Conversely, the direct transform can be expressed as the complex conjugate of the inverse transform applied to an N multiple of the complex conjugate of the original.

This symmetry of the discrete Fourier transform is very significant for practical computation because it allows us to compute the direct and the inverse transforms by the same procedure.

Example 6. A sequence with unit first term and the other terms zero we name δ_i. For this sequence $\delta_0 = 1$, $\delta_i = 0$, $i = 1, 2, \ldots, N - 1$. The image of the sequence δ_i is then given by the defining relation (4-1) as $X_k = 1$, $k = 0, 1, \ldots, N - 1$. Hence we can write $\delta_i \triangleq 1$.

Example 7. The sequence $x_i = 1$, $i = 0, 1, \ldots, N - 1$, has the image $X_0 = N$, $X_k = 0$, $k = 1, 2, \ldots, N - 1$. Using the sequence δ_i introduced in example 6, we can write $1 \triangleq N\delta_k$.

Theorem 3. Linearity of the transform

If ${}_1x_i \mathrel{\hat{=}} {}_1X_k$ *and* ${}_2x_i \mathrel{\hat{=}} {}_2X_k$, *then*

(4-16) $\quad a_1x_i + b_2x_i \mathrel{\hat{=}} a_1X_k + b_2X_k.$

This property follows directly from the defining relations.

Theorem 4: Transform of reversed sequences

As a reversed sequence to the sequence $x_0, x_1, x_2, \ldots, x_{N-1}$ *we define a sequence* $x_0, x_{N-1}, x_{N-2}, \ldots, x_2, x_1$, *i.e. a sequence with general term* x_{N-i} *for which* (4-12) *implies that* $x_{N-i} = x_{-i}$. From the defining relations for the discrete Fourier transform we get

(4-17) $\quad \mathscr{D}\{x_{-i}\} = X_{-k},$

and

(4-18) $\quad \mathscr{D}^{-1}\{X_{-k}\} = x_{-i}.$

The image of the reversed sequence is equal to the reversed image.

Theorem 5: Transforms of an even and an odd sequence

Define an even sequence as one with the property $x_{-i} = x_i$ *and an odd sequence as one where* $x_{-i} = -x_i$. *The image of an even sequence is an even sequence:*

(4-19) $\quad X_{-k} = X_k;$

the image of an odd sequence is an odd sequence:

(4-20) $\quad X_{-k} = -X_k.$

On splitting a sequence x_i *into an even part* $x_i^{\text{even}} = (x_i + x_{-i})/2$ *and an odd part* $x_i^{\text{odd}} = (x_i - x_{-i})/2$, *we get for their images*

(4-21) $$\mathscr{D}\{x_i^{\text{even}}\} = \frac{X_k + X_{-k}}{2},$$

(4-22) $$\mathscr{D}\{x_i^{\text{odd}}\} = \frac{X_k - X_{-k}}{2}.$$

The assertion results from relation (4-17) for the image of the reversed sequence.

Theorem 6: Transform of a complex conjugate sequence

The image of a complex conjugate sequence x_i^* *is equal to a complex conjugate reversed image* X^*_{-k} *of a sequence* x_i. *The original to a complex conjugate image is the reversed complex conjugate sequence* x^*_{-i}:

(4-23) $\mathscr{D}\{x_i^*\} \quad = X^*_{-k}$,

(4-24) $\mathscr{D}^{-1}\{X_k^*\} = x^*_{-i}$.

The assertion follows directly from the defining relations.

On splitting a complex sequence x_i into real and imaginary parts

$$\operatorname{Re} x_i = \frac{1}{2}(x_i + x_i^*),$$

$$\operatorname{Im} x_i = \frac{1}{2j}(x_i - x_i^*),$$

theorem 6 implies

$$\operatorname{Re} x_i \;\hat{=}\; \frac{1}{2}(X_k + X^*_{-k}),$$

$$\operatorname{Im} x_i \;\hat{=}\; \frac{1}{2j}(X_k - X^*_{-k}).$$

Theorem 7: Properties of the transform of a real sequence

1. *If* x_i *is a real sequence then its image fulfils the condition*

(4-25) $X_k^* = X_{-k}$.

This relation follows from the theorem on the image of a complex conjugate sequence, remembering that for a real sequence $x_i = x_i^*$.

2. Using $X_k = U_k + jV_k$ then from expression (4-25) it follows that the real part of the image of a real sequence is even,

(4-26) $U_{-k} = U_k$,

and the imaginary part odd,

(4-27) $V_{-k} = -V_k$.

From the properties given for the image of a real sequence the following important facts result:

3. (a) *If* N *is even, then for the determination of* X_k *it is sufficient to know, because of* (4-20), *only these terms:*

(4-28) U_k, $\quad k = 0, 1, \ldots, N/2$,

(4-29) V_k, $\quad k = 1, 2, \ldots, N/2 - 1$,

the other values being determined in the following way:

(4-30) $$U_{N-k} = U_k, \quad V_{N-k} = -V_k, \qquad k = 1, 2, \ldots, N/2 - 1,$$

while X_0 and $X_{N/2}$ are always real and to them the following applies

(4-31) $$X_0 = U_0 = \sum_{i=0}^{N-1} x_i,$$

(4-32) $$X_{N/2} = U_{N/2} = \sum_{i=0}^{N-1} (-1)^i x_i.$$

(b) *If N is odd, then for the determination of X_k it is sufficient to know only*

(4-33) $$U_k, \quad V_k, \qquad k = 0, 1, \ldots, (N-1)/2,$$

and the other values are determined by the relations

(4-34) $$U_{N-k} = U_k, \quad V_{N-k} = -V_k, \qquad k = 1, 2, \ldots, (N-1)/2.$$

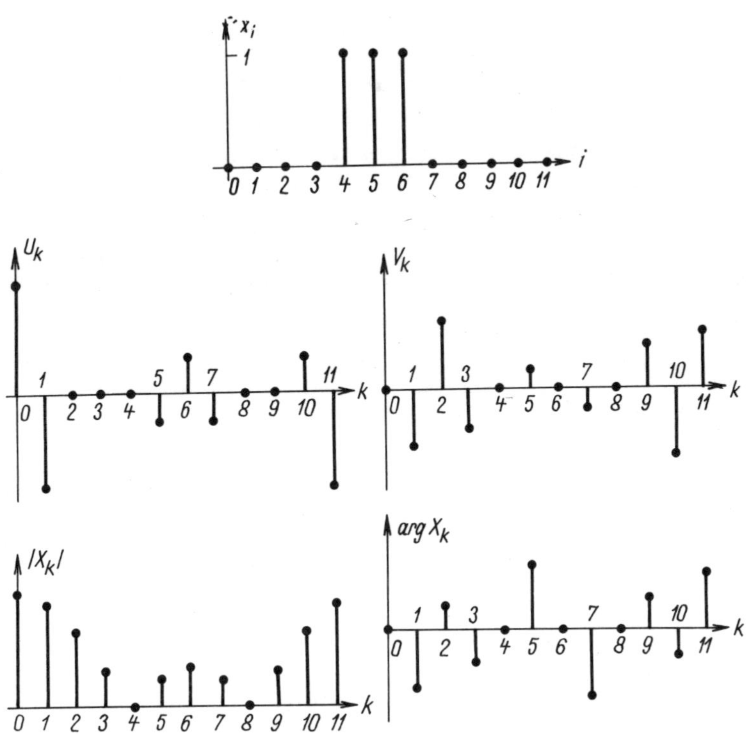

7. The real sequence from example 8 and its discrete Fourier transform

The term X_0 is always real and obeys

$$(4\text{-}35)\qquad X_0 = U_0 = \sum_{i=0}^{N-1} x_i .$$

4. *If we split x_i into an even part x_i^{even} and an odd part x_i^{odd}*

$$(4\text{-}36)\qquad x_i^{\text{even}} = \tfrac{1}{2}(x_i + x_{-i}),$$

$$(4\text{-}37)\qquad x_i^{\text{odd}} = \tfrac{1}{2}(x_i - x_{-i}),$$

and use the notation $X_k = U_k + jV_k$, then

$$(4\text{-}38)\qquad x_i^{\text{even}} \mathrel{\hat{=}} \tfrac{1}{2}(X_k + X_k^*) = U_k,$$

$$(4\text{-}39)\qquad x_i^{\text{odd}} \mathrel{\hat{=}} \tfrac{1}{2}(X_k - X_k^*) = jV_k .$$

For a real sequence, the image of its even part is the real part of the image and the image of its odd part is the imaginary part of the image multiplied by j.

This property follows from the theorem on the image of a reversed sequence and from the property of the image of a real function (4-25) applied to (4-36) and (4-37).

Example 8. The real sequence

$$x_i = \begin{cases} 1, & i = 4, 5, 6, \\ 0, & i = 0, 1, 2, 3, 7, \ldots, 11, \end{cases}$$

is shown in Figure 7. In this figure also the real part U_k, imaginary part V_k, modulus and argument of its transform are illustrated.

Theorem 8: Properties of the transform of a complex sequence

Let us use for a complex sequence $z_i = x_i + jy_i$ and for its image $z_i \mathrel{\hat{=}} Z_k = U_k + jV_k$, where x_i, y_i, U_k and V_k are real sequences.

1. *For the image of the real part x_i*

$$(4\text{-}40)\qquad x_i \mathrel{\hat{=}} (Z_k + Z_{-k}^*)/2$$

and for the image of the imaginary part

$$(4\text{-}41)\qquad y_i \mathrel{\hat{=}} (Z_k - Z_{-k}^*)/2j.$$

Both these relations follow from relation (4-23) if we write $x_i = (z_i + z_i^*)/2$ and $y_i = (z_i - z_i^*)/2j$.

If we split U_k and V_k into even and odd parts, $U_k = U_k^{\text{even}} + U_k^{\text{odd}}$, $V_k = V_k^{\text{even}} + V_k^{\text{odd}}$, then for the images of real and imaginary parts

$$(4\text{-}42)\qquad x_i \mathrel{\hat{=}} U_k^{\text{even}} + jV_k^{\text{odd}},$$

$$(4\text{-}43)\qquad y_i \mathrel{\hat{=}} V_k^{\text{even}} - jU_k^{\text{odd}}.$$

We get this kind of expression from relations (4-40) or (4-41) by substituting $Z_k = U_k + jV_k$ and expanding U_k and V_k into even and odd parts.

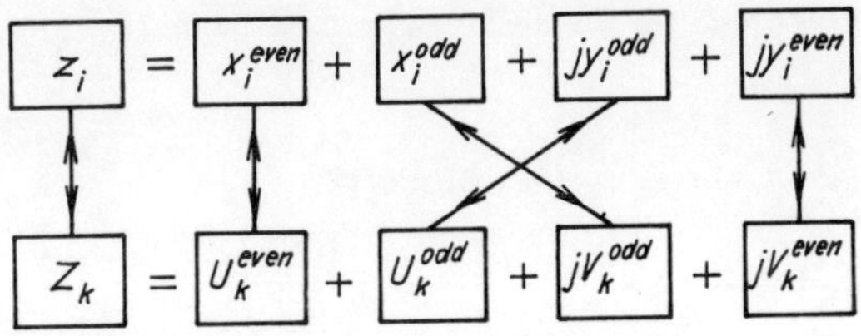

8. Relationship between even and odd parts of an original and its transform

2. *For the real and imaginary parts of the image it holds*

(4-44) $\quad U_k = \mathscr{D}\{x_i^{\text{even}} + \text{j}y_i^{\text{odd}}\},$

(4-45) $\quad V_k = \mathscr{D}\{y_i^{\text{even}} - \text{j}x_i^{\text{odd}}\}.$

We get these relations by applying (4-24) to a sequence z_i and writing $U_k = (Z_k + Z_k^*)/2$ or $V_k = (Z_k - Z_k^*)/2\text{j}$.

Mutual assignments of particular components of the original and image sequences are summarized graphically in Figure 8.

4.1.3 Matrix notation of the discrete Fourier transform

Theorem 9: Matrix expression of a transform

If we write a sequence x_i *as a column vector*

$$^T\mathbf{x} = (x_0, x_1, \ldots, x_{N-1})$$

and its image X_k *likewise as a column vector*

$$^T\mathbf{x} = (X_0, X_1, X_2, \ldots, X_{N-1}),$$

then relation (4-1) *for the direct transform may be written in matrix form as*

(4-46) $\quad \mathbf{X} = \mathbf{W}\mathbf{x}.$

The matrix $\mathbf{W}$ *is a square matrix of order* N,

(4-47)
$$\mathbf{W} = \begin{pmatrix} W^0, & W^0, & W^0, & \ldots, & W^0 \\ W^0, & W^1, & W^2, & \ldots, & W^{N-1} \\ W^0, & W^2, & W^4, & \ldots, & W^{2(N-1)} \\ \vdots & & & & \\ W^0, & W^{N-1}, & W^{2(N-1)}, & \ldots, & W^{(N-1)(N-1)} \end{pmatrix}$$

and for its general term

(4-48) $\quad w_{i,k} = W^{ik}, \quad W = \text{e}^{-\text{j}2\pi/N}, \qquad i = 0, 1, \ldots, N-1,$
$\qquad\qquad k = 0, 1, \ldots, N-1.$†)

†) For the sake of consistency with indexing terms of a sequence we shall always index matrix elements this way.

The inverse transform (4-2) *may also be expressed in matrix language*

(4-49) $\quad \mathbf{x} = \mathbf{W}^{-1}\mathbf{X},$

where $\mathbf{W}^{-1}$ *is the inverse matrix to the matrix* $\mathbf{W}$.

The matrix $\mathbf{W}$ possesses a number of properties (see e.g.[1]), some of which will be presented here.

Theorem 10: Properties of the matrix W

1. *The matrix* $\mathbf{W}$ *is regular and symmetrical:*

(4-50) $\quad {}^{T}\mathbf{W} = \mathbf{W}, \qquad {}^{T}(\mathbf{W}^{-1}) = \mathbf{W}^{-1}.$

2. *The matrix* $\mathbf{W}^{-1}$ *is equal to the complex conjugate matrix* $\mathbf{W}$ *divided by N:*

(4-51) $\quad \mathbf{W}^{-1} = (1/N)\,\mathbf{W}^{*}.$

3. *The matrix* $\mathbf{W}$ *is unitary except for a factor* $1/N$, *i.e.*

(4-52) $\quad (T_{\mathbf{W}})^{*} = (1/n)\,\mathbf{W}^{-1}$

4. *The matrix* $\mathbf{W}$ *is a permutation periodic matrix of 4-th order with the following properties:*

(4-53) $\quad \mathbf{W}^{4} = N^{2}\mathbf{E},$

(4-54) $\quad \mathbf{W}^{2} = N\mathbf{P},$

where $\mathbf{E}$ *is a unit matrix of order N and* $\mathbf{P}$ *is a permutation matrix of order N with general term*

(4-55) $$p_{ik} = \begin{cases} 1, & i = k = 0, \\ 1, & i + k = N, \\ 0, & \text{the other } i,\, k. \end{cases}$$

We first prove relation (4-54). If we use v_{ik} for the general term of the matrix $\mathbf{V} = \mathbf{W}\,.\,\mathbf{W} = \mathbf{W}^{2}$, then it obeys

$$v_{ik} = \sum_{\nu=0}^{N-1} w_{ik} w_{\nu k} = \sum_{\nu=0}^{N-1} \mathrm{e}^{-\mathrm{j}\nu(i+k)\,2\pi/N} = \begin{cases} N, & i = k = 0, \\ N, & i + k = N, \\ 0, & \text{the other } i,\, k, \end{cases}$$

and thence $v_{ik} = Np_{ik}$. Relation (4-53) follows from relation (4-54), given $\mathbf{P}^{2} = \mathbf{E}$.

5. *The matrix* $\mathbf{W}$ *has eigenvalues* [63]

$$\lambda_1 = \sqrt{N}, \quad \lambda_2 = -\sqrt{N}, \quad \lambda_3 = -\mathrm{j}\sqrt{N}, \quad \lambda_4 = \mathrm{j}\sqrt{N}$$

with multiples m_ν, $\nu = 1, 2, 3, 4$, given in Table 3. In the table the values of the determinant det $\mathbf{W}$ of the matrix $\mathbf{W}$ are also shown.

Table 3.

N	m_1	m_2	m_3	m_4	det **W**
$4n$	$n+1$	n	n	$n-1$	$-\mathrm{j}(-1)^n N^{N/2}$
$4n+1$	$n+1$	n	n	n	$(-1)^n N^{N/2}$
$4n+2$	$n+1$	$n+1$	n	n	$-N^{N/2}$
$4n+3$	$n+1$	$n+1$	$n+1$	n	$\mathrm{j}(-1)^n N^{N/2}$

6. *If we use* $\mathbf{u}_1$ *for a vector constituting a row of the matrix* **W** *for* $i = 1$,

$$\mathbf{u}_1 = (W^0, W^1, W^2, \dots, W^{N-1}),$$

then every i-th row can be expressed as a vector $\mathbf{u}_i$, *elements of which are i-th powers of elements constituting the vector* $\mathbf{u}_1$:

(4-56) $\quad \mathbf{u}_i = (W^0, W^i, W^{2i}, \dots, W^{(N-1)i}).$

7. *The elements* W^k, $k = 0, 1, \dots, N-1$, *of the vector* $\mathbf{u}_1$ *are roots of the equation*

(4-57) $\quad \mathbf{z}^N - 1 = 0$

and further

$$\sum_{k=0}^{N-1} W^{ik} = \begin{cases} N, & i = 0, \\ 0, & i = 1, 2, \dots, N-1, \end{cases}$$

$$\prod_{k=0}^{N-1} W^{ik} = -(-1)^{iN}, \qquad i = 0, 1, \dots, N-1.$$

The numbers $W^k = \cos(k \,.\, 2\pi/N) - \mathrm{j}\sin(k \,.\, 2\pi/N)$, $k = 0, 1, \dots, N-1$, *lie in the complex plane* $\mathbf{z}$ *on the unit circle and divide into N equal parts beginning with a root* $W^0 = 1$. *The complex numbers* W^k *can be represented as a set of vectors in the complex plane with a common origin or as a set of free vectors which link one to another and create a regular N-gon.*

4.1.4 Geometrical interpretation of the discrete Fourier transform

A sequence of complex numbers X_k, $k = 0, 1, \dots, N-1$, can be interpreted as a general set of plane vectors with a common origin. The defining relation for X_k may then be considered as the expansion of each vector X_k, $k = 0, 1, \dots, N-1$, in terms of the set of N vectors $(W^k)^i$, $i = 0, 1, \dots, N-1$:

(4-58) $\quad X_k = \sum_{i=0}^{N-1} x_i (W^k)^i.$

The quantities x_i have the meaning of coefficients of this expansion and are given by the inverse transform of the sequence of numbers X_k,

$$x_i = \frac{1}{N} \sum_{k=0}^{N-1} X_k W^{-ik}.$$

This way of expanding a common set of N vectors into N sets is known in the theory of multiphase currents as an expansion into so-called symmetric components, introduced by C. L. Fortescue. The relationship of this expansion to the finite Fourier series was demonstrated by E. A. Guillemin [48] and V. Klíma [55].

4.2 Other properties of discrete Fourier transforms

We present in this section other properties of the discrete Fourier transform, namely properties of images of the most frequently used operations with sequences, such as their shifting, periodic and discrete convolution and correlation, product and scalar product.

To be able to express those properties we introduce first the *product* of two *sequences* of the same length x_i and y_i, $i = 0, 1, \ldots, N-1$, as the sequence with general term $x_i y_i$, $i = 0, 1, \ldots, N-1$. Furthermore we introduce the *scalar product* of two *sequences* of the same length, as a number determined by the relation

$$\sum_{i=0}^{N-1} x_i y_i.$$

4.2.1 Discrete Fourier transform of shifted sequences

We shall further assume that sequences x_i and X_i are periodic and introduce the shift of a sequence

$$y_i = x_{i-\nu}, \qquad \nu = \pm 1, \pm 2, \ldots, \pm(N-1), \tag{4-59}$$

$$X_k = X_{k-\mu}, \qquad \mu = \pm 1, \pm 2, \ldots, \pm(N-1). \tag{4-60}$$

It is obvious, because of the periodicity of x_i and X_k, that a shift of ν is equivalent to a shift of $\nu - N$ and a shift of μ is equivalent to a shift of $\mu - N$.

Examples of shifted sequences are shown in Figure 9.

Theorem 11: Transform of a shifted sequence

The image of a sequence x_i, shifted by ν, is equal to the image of the sequence x_i multiplied by $e^{-j\nu k 2\pi/N}$.

$$\mathscr{D}\{x_{i-\nu}\} = X_k\, e^{-j\nu k 2\pi/N}. \tag{4-61}$$

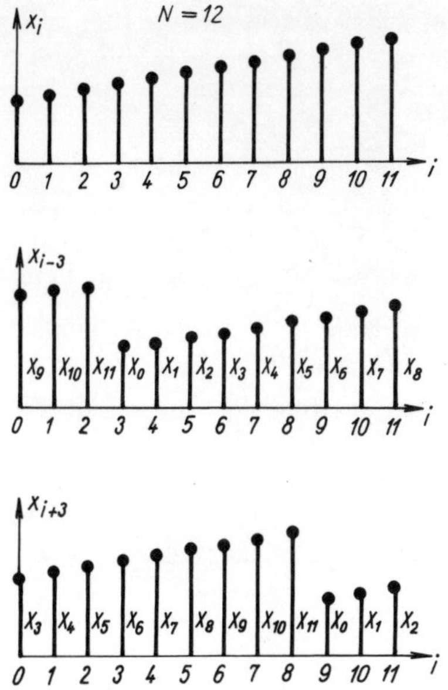

9. Shift of a sequence

From the defining relation for X_k we get, on substituting $i = \mu + N$,

$$\sum_{i=0}^{N-1} x_{i-\nu}\, e^{-jik2\pi/N} = \sum_{\mu=-\nu}^{N-1-\nu} x_\mu\, e^{-j(\mu+\nu)k2\pi/N} =$$

$$= e^{j\nu k2\pi/N}\Big(\sum_{\mu=0}^{N-1-\nu} x_\mu\, e^{-j\mu k2\pi/N} + \sum_{\mu=-\nu}^{-1} x_\mu\, e^{-j\mu k2\pi/N}\Big).$$

After introducing the substitution $i = \mu + N$ in the last sum we get

$$\sum_{\mu=-\nu}^{-1} x_\mu\, e^{j\mu k2\pi/N} = \sum_{N-\nu}^{N-1} x_{i+N}\, e^{jik2\pi/N}.$$

Considering also a periodicity x_i we get

$$\sum_{i=0}^{N-1} x_{i-\nu}\, e^{-jik2\pi/N} = e^{-j\nu k2\pi/N} \sum_{i=0}^{N-1} x_i\, e^{-jik2\pi/N}$$

and the assertion of theorem 11 is proved.

Example 9. A sequence $x_i = \delta_{i-\nu}$ made by shifting the sequence δ_i from example 6, i.e. a sequence $x_i = 1$, $i = \nu$, $x_i = 0$, $i = 0, 1, \ldots, \nu - 1, \nu + 1, \ldots, N - 1$, has, according to example 6 and by theorem 11, an image $X_k = e^{-j\nu k2\pi/N}$, $k = 0, 1, \ldots, N - 1$, and hence we can write $\delta_{i-\nu} \mathrel{\hat{=}} e^{-j\nu k2\pi/N}$.

By analogy the following theorem on the shift in an image can also be proved.

Theorem 11a: Inverse transform of a shifted sequence

The original of a shifted image $X_{k-\mu}$ *is a sequence* x_i *multiplied by the exponential sequence* $e^{j\mu i 2\pi/N}$:

(4-62) $\mathscr{D}^{-1}\{X_{k-\mu}\} = x_i\, e^{j\mu i 2\pi/N}$.

Both these assertions can, of course, be interpreted conversely. However, as they possess some practical meaning, we present them here as independent theorems.

Theorem 12: Transform of a sequence multiplied by an exponential sequence

The image of a sequence x_i *multiplied by an exponential sequence* $e^{j\mu i 2\pi/N}$ *is equal to the image shifted by* μ *of the sequence* x_i:

(4-63) $\mathscr{D}\{x_i\, e^{j\mu i 2\pi/N}\} = X_{k-\mu}$.

Example 10. Values of a function

$$x(t) = e^{j\varkappa 2\pi t/P}, \qquad \varkappa = 1, 2, \ldots, N-1$$

for $t = iT$, $T = P/N$, $i = 0, 1, \ldots, N-1$, define a sequence $x_i = x(iT) = e^{ji\varkappa 2\pi/N}$. We obtain for the image, from (4-1), $X_k = N$, $k = \varkappa$, $X_k = 0$, $k \neq \varkappa$. Using the sequence δ_i from example 6 to express the image, we can write $X_k = N\delta_{k-\varkappa}$. If we compare the expression obtained with the Fourier transform of the function $e^{j\varkappa 2\pi t/P} \mathrel{\hat{=}} 2\pi\delta(\omega - \varkappa\,.\,2\pi/P)$, we see a direct analogy that will be mentioned in § 5.1.1.

Example 11. The function $x(t) = \cos(\varkappa\,.\,2\pi t/P)$, $\varkappa = 1, 2, \ldots, N/2 - 1$, N even or $\varkappa = 1, 2, \ldots, (N-1)/2$, N odd defines by its values for $t = iT$, $T = P/N$, a sequence $x_i = \cos(i\varkappa\,.\,2\pi/N)$. Using the linearity of the transform and example 10, we get for the image of a cosine sequence $X_k = N(\delta_{k-\varkappa} + \delta_{k+\varkappa})/2$. By analogy we get for the image of a sine sequence $x_i = \sin(i\varkappa\,.\,2\pi N)$ the expression $X_k = N(\delta_{k-\varkappa} - \delta_{k+\varkappa})/2j$.

Theorem 12a: Inverse transform of a sequence multiplied by an exponential sequence

The original of a sequence X_k *multiplied by an exponential sequence* $e^{j\nu k 2\pi/N}$ *is a sequence* x_i *shifted by* $-\nu$:

(4-64) $\mathscr{D}^{-1}\{X_k\, e^{j\nu k 2\pi/N}\} = x_{i+\nu}$.

From the theorems presented another important conclusion follows, *that the image of an original sequence and the image of a shifted sequence have the same absolute value.*

4.2.2 Discrete Fourier transform of a periodic convolution and a product of images

First we define a *discrete periodic convolution* of sequences of the same length $x_i \triangleq X_k$ and $y_i \triangleq Y_k$, $i = 0, 1, \ldots, N-1$, which we consider to be periodic, by means of the relation

$$z_i = \sum_{\nu=0}^{N-1} x_\nu y_{i-\nu} = \sum_{\nu=0}^{N-1} y_\nu x_{i-\nu}, \qquad i = 0, 1, \ldots, N-1. \tag{4-65}$$

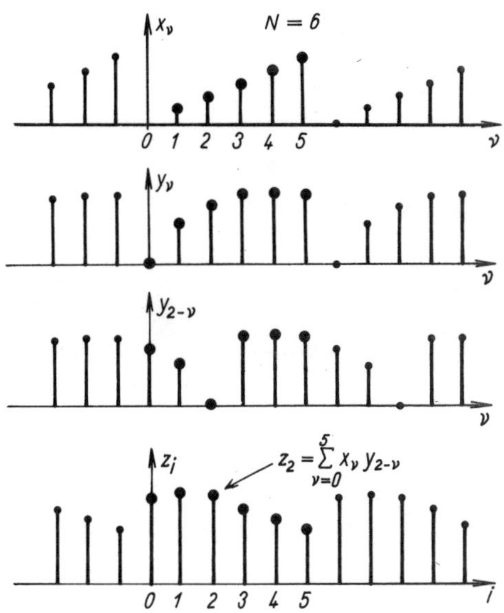

10. The discrete periodic convolution of sequences x_i and y_i

Successive creation of the convolution sum is illustrated for $i = 2$ in Figure 10. Symbolically we shall write this convolution

$$z_i = x_i \circledast y_i, \tag{4-66}$$

to distinguish it from the usual (non-periodic) discrete convolution that will be introduced in § 7.1.1. In that section we shall also show a mutual relationship between both kinds of convolution. The following theorem applies for an image of this convolution.

Theorem 13: Transform of a periodic convolution

The image of a discrete periodic convolution z_i of sequences $x_i \triangleq X_k$ and $y_i \triangleq Y_k$ is equal to the product of their images

$$\mathcal{D}\{z_i\} = \mathcal{D}\{x_i \circledast y_i\} = X_k Y_k, \qquad k = 0, 1, \ldots, N-1. \tag{4-67}$$

Carrying out the transform of a convolution sum

$$\mathscr{D}\{z_i\} = \sum_{i=0}^{N-1} \sum_{\nu=0}^{N-1} x_\nu y_{i-\nu}\, \mathrm{e}^{-\mathrm{j}ik2\pi/N}$$

and then using theorem 11 on the image of a shifted sequence, we obtain directly the following result

$$\mathscr{D}\{z_i\} = \sum_{\nu=0}^{N-1} x_\nu Y_k\, \mathrm{e}^{-\mathrm{j}\nu k2\pi/N} = Y_k \mathscr{D}\{x_\nu\} = X_k Y_k.$$

In the reverse direction, the theorem can be proved using the defining relation for $\mathscr{D}^{-1}$

$$z_i = \mathscr{D}^{-1}\{X_k Y_k\} = \frac{1}{N} \sum_{k=0}^{N-1} X_k Y_k\, \mathrm{e}^{\mathrm{j}ik2\pi/N}$$

and theorem 12a, thus obtaining

$$z_i = \frac{1}{N} \sum_{k=0}^{N-1} X_k \sum_{\nu=0}^{N-1} y_{\nu+i}\, \mathrm{e}^{-\mathrm{j}\nu k2\pi/N} =$$

$$= \sum_{\nu=0}^{N-1} y_{\nu+i} \frac{1}{N} \sum_{k=0}^{N-1} X_k\, \mathrm{e}^{-\mathrm{j}\nu k2\pi/N} = \sum_{\nu=0}^{N-1} y_{\nu+i} x_{-\nu}.$$

Using the periodicity of sequences and introducing $\mu = N - \nu$, we get

$$z_i = \sum_{\mu=1}^{N} y_{i-(\mu-N)} x_{\mu-N} = \sum_{\mu=1}^{N-1} y_{i-(\mu-N)} x_{\mu-N} + y_i x_0 =$$

$$= \sum_{\mu=1}^{N-1} y_{i-\mu} x_\mu + y_i x_0 = \sum_{\mu=0}^{N-1} y_{i-\mu} x_\mu.$$

Hence, the theorem is proved in the reverse direction and can, therefore, also be expressed in the following manner:

Theorem 13a: Inverse transform of a product of image sequences

The original of a product of two image sequences is a discrete periodic convolution of sequences of originals.

Example 12. Relation (4-61) applies to the image of a shifted sequence. If we use theorem 13a, we may take this relation as the image of a convolution. As a result, a shifted sequence $x_{i-\nu}$ can be interpreted as the convolution of the original sequence x_i with a sequence $\delta_{i-\nu}$ and thus we can write $x_{i-\nu} = x_i \circledast \delta_{i-\nu}$.

A discrete periodic convolution can also be introduced for the image sequences:

$$\sum_{\varkappa=0}^{N-1} X_\varkappa Y_{k-\varkappa} = \sum_{\varkappa=0}^{N-1} Y_\varkappa X_{k-\varkappa} = X_k \circledast Y_k = Y_k \circledast X_k.$$

In this case theorem 14 holds.

Theorem 14: Inverse transform of a periodic convolution of images

The original of a periodic convolution of images $X_k \triangleq x_i$, $Y_k \triangleq y_i$ *is an N-multiple of a product of originals*

(4-68) $$\mathscr{D}^{-1}\{X_k \circledast Y_k\} = N x_i y_i, \qquad i = 0, 1, \ldots, N-1.$$

For the proof of this theorem we expand the inverse transform as follows:

$$\mathscr{D}^{-1}\{X_k \circledast Y_k\} = \mathscr{D}^{-1}\left\{\sum_{\varkappa=0}^{N-1} X_\varkappa Y_{k-\varkappa}\right\} =$$

$$= \sum_{\varkappa=0}^{N-1} X_\varkappa \mathscr{D}^{-1}\{Y_{k-\varkappa}\}.$$

Using theorem 11a for the shift of an image sequence we obtain the following assertion

$$\mathscr{D}^{-1}\{X_k \circledast Y_k\} = \sum_{\varkappa=0}^{N-1} X_\varkappa y_i \, e^{j\varkappa i 2\pi/N} =$$

$$= y_i \sum_{\varkappa=0}^{N-1} X_\varkappa \, e^{j\varkappa i 2\pi/N} = N x_i y_i.$$

This theorem may also be expressed conversely:

Theorem 14a: Transform of a product of sequences

The image of a product of sequences x_i *and* y_i *is equal to a discrete periodic convolution of images* X_k *and* Y_k *of these sequences divided by* N*:*

(4-69) $$\mathscr{D}\{x_i y_i\} = \frac{1}{N} X_k \circledast Y_k.$$

4.2.3 Discrete periodic correlation

We may call the expression

(4-70) $$\varrho_i(x_\nu, y_\nu) = \sum_{\nu=0}^{N-1} x_\nu y_{\nu+i}, \qquad i = 0, 1, \ldots, N-1,$$

the discrete periodic correlation of a sequence x_i and a sequence y_i because of its similarity to the discrete correlation. Further we just use the term correlation, where there is no possibility of misunderstanding.

(A) Relationship between the correlation and the correlation in reverse order

The correlation defined by relation (4-70) is not commutative, and, therefore, a correlation of sequences in reverse order defines another sequence

(4-71) $$\varrho_i(x_\nu, y_\nu) = \sum_{\nu=0}^{N-1} y_\nu x_{\nu+i}, \qquad i = 0, 1, \ldots, N-1.$$

If we replace $\nu + i = \mu$ in relation (4-70), we obtain

$$\varrho_i(x_\nu, y_\nu) = \sum_{\mu=i}^{i+N-1} x_{\mu-i} y_\mu = \sum_{\mu=i}^{N-1} x_{\mu-i} y_\mu + \\ + \sum_{\mu=N}^{i+N-1} x_{\mu-i} y_\mu .$$

On substituting $\varkappa = \mu - N$ in the second sum we arrive at an expression

$$\varrho_i(x_\nu, y_\nu) = \sum_{\mu=i}^{N-1} x_{\mu-i} y_\mu + \sum_{\varkappa=0}^{i-1} x_{\varkappa+N-i} y_{\varkappa+N} = \\ = \sum_{\mu=0}^{N-1} y_\mu x_{\mu-i} = \varrho_{-i}(y_\nu, x_\nu).$$

Now, using the terminology from § 4.1.2, where a sequence with negatively taken indices was called the reversed sequence, one can express the relation as follows:

The correlation of sequences in reverse order is equal to the reversed correlation of sequences in the original order:

(4-72) $\varrho_i(y_\nu, x_\nu) = \varrho_{-i}(x_\nu, y_\nu).$

Both cases of correlation are illustrated in Figure 11.

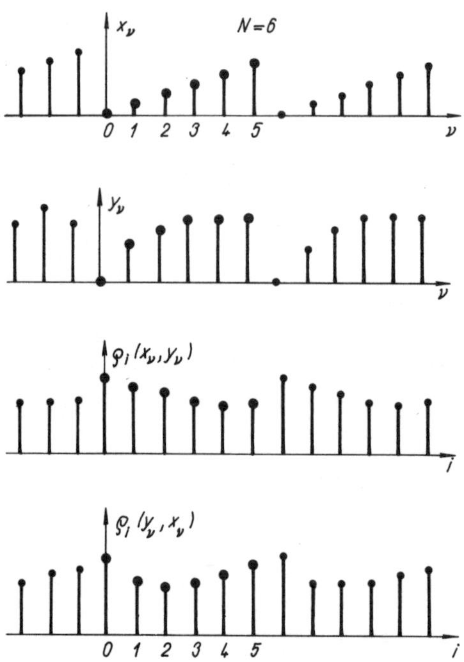

11. The discrete periodic correlation $\varrho_i(x_\nu, y_\nu)$ and $\varrho_i(y_\nu, x_\nu)$

(B) Relationship of the correlation to a convolution

The expression for the correlation can be adjusted by substituting $\mu = N - v$ into the convolution

$$\varrho_i(x_v, y_v) = \sum_{v=0}^{N-1} x_v y_{v+i} = \sum_{\mu=1}^{N} x_{N-\mu} y_{i+N-\mu} =$$
$$= \sum_{\mu=1}^{N} x_{-\mu} y_{i-\mu} = \sum_{\mu=0}^{N-1} x_{-\mu} y_{i-\mu} = x_{-\mu} \circledast y_\mu .$$

The correlation $\varrho_i(x_v, y_v)$ is equal to the convolution of the reversed sequence x_{-v} with the sequence y_v:

(4-73a) $\varrho_i(x_v, y_v) = x_{-v} \circledast y_v .$

From this relation we get for the reversed sequence x_{-v}

(4-73b) $x_v \circledast y_v = \varrho_i(x_{-v}, y_v).$

The convolution $x_v \circledast y_v$ is equal to the correlation of the reversed sequence x_{-v} with the sequence y_v.

Theorem 15: Transform of a periodic correlation

The image of a discrete periodic correlation $\varrho_i(x, y)$ of a sequence $x_i \triangleq X_k$ and a sequence $y_i \triangleq Y_k$ equals the product of the reversed image of the first sequence with the image of the second sequence:

(4-74) $\mathscr{D}\{\varrho_i(x_v, y_v)\} = X_{-k} Y_k .$

For the image of the correlation of sequences x_i and y_i in reverse order it is true that

(4-75) $\mathscr{D}\{\varrho_i(y_v, x_v)\} = \mathscr{D}\{\varrho_{-i}(x_v, y_v)\} = X_k Y_{-k} .$

For proof we expand the transforms of the correlation:

$$\mathscr{D}\{\varrho_i(x_v, y_v)\} = \mathscr{D}\{\sum_{v=0}^{N-1} x_v y_{v+i}\} = \sum_{v=0}^{N-1} x_v \mathscr{D}\{y_{i+v}\}.$$

Theorem 11 on the shift enables us to arrive at the assertion

$$\mathscr{D}\{\varrho_i(x, y)\} = Y_k \sum_{v=0}^{N=1} x_v \, e^{jvk\,2\pi/N} = Y_k X_{-k} .$$

By applying theorem 4 on the image of a reversed sequence to relation (4-73) we obtain (4-75) for the image of the correlation of sequences x_i and y_i in reverse order.

In the case that both sequences are the same, $y_i = x_i$, then for the image of the periodic correlation (sometimes called also the periodic autocorrelation) the following holds

$$\mathscr{D}\{\varrho_i(x_\nu, x_\nu)\} = X_k X_{-k}. \tag{4-76}$$

Since both sequences are real,

$$\mathscr{D}\{\varrho_i(x_\nu, y_\nu)\} = X_k^* Y_k, \tag{4-77}$$

$$\mathscr{D}\{\varrho_i(x_\nu, x_\nu)\} = |X_k|^2. \tag{4-78}$$

The discrete periodic correlation can also be defined for sequences representing images:

$$\varrho_k(X_\varkappa, Y_\varkappa) = \sum_{\varkappa=0}^{N-1} X_\varkappa Y_{\varkappa+k}, \tag{4-79}$$

$$\varrho_k(Y_\varkappa, X_\varkappa) = \sum_{\varkappa=0}^{N-1} Y_\varkappa X_{\varkappa+k}, \tag{4-80}$$

where, again,

$$\varrho_{-k}(X_\varkappa, Y_\varkappa) = \varrho_k(Y_\varkappa, X_\varkappa). \tag{4-81}$$

By analogy with the proof of theorem 15, we can now prove an analogous theorem for the inverse transform of the sequence $\varrho_k(X_\varkappa, Y_\varkappa)$.

Theorem 16: Inverse transform of the periodic correlation

The original of the discrete periodic correlation $\varrho_k(X_\varkappa, Y_\varkappa)$ of images $X_k \triangleq x_i$ and $Y_k \triangleq y_i$ is an N-multiple product of the reversed sequence x_{-i} with the sequence y_i:

$$\mathscr{D}^{-1}\{\varrho_k(X_\varkappa, Y_\varkappa)\} = N x_{-i} y_i. \tag{4-82}$$

The original of a correlation of the same sequence in reverse order is a reversed sequence

$$\mathscr{D}^{-1}\{\varrho_k(Y_\varkappa, X_\varkappa)\} = \mathscr{D}^{-1}\{\varrho_{-k}(X_\varkappa, Y_\varkappa)\} = N y_{-i} x_i. \tag{4-83}$$

4.2.4 Properties of a sum of sequence elements and scalar products

From the defining relation for the direct and inverse transforms we get the following two assertions:

Theorem 17: The sum of elements of an original sequence

The sum of elements of an original sequence x_i is equal to the zero element of an image sequence X_k:

$$\sum_{i=0}^{N-1} x_i = X_0. \tag{4-84}$$

Theorem 17a: The sum of elements of an image sequence

The sum of elements of an image sequence equals an N-multiple of the zero term of the original sequence:

$$\sum_{k=0}^{N-1} X_k = N x_0. \qquad (4\text{-}85)$$

Further on we present a theorem on the scalar product of a sequence that is also called Parseval's theorem. The scalar product was introduced at the beginning of § 4.2.

Theorem 18: Equality of the scalar product of an original and an image

The scalar product of a sequence x_i with a sequence y_i is equal to the scalar product of the image of one sequence with a reversed image of the other sequence divided by N:

$$\sum_{i=0}^{N-1} x_i y_i = \frac{1}{N}\sum_{k=0}^{N-1} X_{-k} Y_k = \frac{1}{N}\sum_{k=0}^{N-1} X_k Y_{-k}. \qquad (4\text{-}86)$$

The proof follows from theorem 15 on the image of a correlation after setting $i = 0$ in relation (4-70) or (4-71):

$$\varrho_0(x_\nu, y_\nu) = \varrho_0(y_\nu, x_\nu) = \sum_{\nu=0}^{N-1} x_\nu y_\nu .$$

Now, on applying theorem 17a for the sum of terms of an image sequence to relation (4-74) or (4-75), we get

$$\varrho_0(x_\nu, y_\nu) = \frac{1}{N}\sum_{k=0}^{N-1} X_{-k} Y_k$$

and

$$\varrho_0(y_\nu, x_\nu) = \frac{1}{N}\sum_{k=0}^{N-1} X_k Y_{-k},$$

and the theorem is thus proved.

Using the periodicity of the sequences, it is obvious that

$$\sum_{i=0}^{N-1} x_i y_i = \sum_{i=0}^{N-1} x_{-i} y_{-i} \qquad (4\text{-}87)$$

and

$$\sum_{k=0}^{N-1} X_k Y_k = \sum_{k=0}^{N-1} X_{-k} Y_{-k}. \qquad (4\text{-}88)$$

Relation (4-86) expressing theorem 18 can be modified in various ways. If we introduce instead of x_i a reversed sequence $x_{-i} \triangleq X_{-k}$, we get from (4-86)

$$\sum_{i=0}^{N-1} x_{-i} y_i = \frac{1}{N}\sum_{k=0}^{N-1} X_k Y_k. \qquad (4\text{-}89)$$

Then using theorem 4 on the image of the reversed sequence one gets

$$\sum_{i=0}^{N-1} x_i y_{-i} = \frac{1}{N}\sum_{k=0}^{N-1} X_{-k} Y_{-k} = \frac{1}{N}\sum_{k=0}^{N-1} X_k Y_k,$$

and hence

$$\sum_{k=0}^{N-1} X_k Y_k = N\sum_{i=0}^{N-1} x_i y_{-i} = N\sum_{i=0}^{N-1} x_{-i} y_i. \tag{4-90}$$

If we use instead of x_i a sequence $x_i^* \mathrel{\hat=} X_{-k}^*$, we get from (4-86)

$$\sum_{i=0}^{N-1} x_i^* y_i = \frac{1}{N}\sum_{k=0}^{N-1} X_k^* Y_k \tag{4-91}$$

and by creating complex conjugate values

$$\sum_{i=0}^{N-1} x_i y_i^* = \frac{1}{N}\sum_{k=0}^{N-1} X_k Y_k^*. \tag{4-92}$$

When both the sequences are the same, $x_i = y_i$, we get the following relations: from (4-86)

$$\sum_{i=0}^{N-1} x_i^2 = \frac{1}{N}\sum_{k=0}^{N-1} X_{-k} X_k, \tag{4-93}$$

$$\sum_{k=0}^{N-1} X_k^2 = N\sum_{i=0}^{N-1} x_i x_{-i} \tag{4-94}$$

and from (4-91)

$$\sum_{i=0}^{N-1} |x_i|^2 = \frac{1}{N}\sum_{k=0}^{N-1} |X_k|^2. \tag{4-95}$$

For the real x_i it then follows that

$$\sum_{i=0}^{N-1} x_i^2 = \frac{1}{N}\sum_{k=0}^{N-1} |X_k|^2. \tag{4-96}$$

4.3 Discrete Fourier transform of modified sequences

In this section we present some properties of the discrete Fourier transform of sequences following certain modifications of the original sequence. Practical applications of these properties will be shown in later chapters and particularly in Chapter 6.

4.3.1 Discrete Fourier transform of sequences padded with zero terms

From a sequence $x_i \mathrel{\hat=} X_k$, $i = 0, 1, \ldots, M-1$, we construct a padded sequence $y_i \mathrel{\hat=} Y_k$, $i = 0, 1, \ldots, N-1$, $N > M$, by adding $N - M$ zero terms after its last term. For such a padded sequence the following is true:

$$(4\text{-}97)\qquad y_i = \begin{cases} x_i, & i = 0, 1, \dots, M-1, \\ 0, & i = M, M+1, \dots, N-1. \end{cases}$$

For a sequence padded in this way we again assume that it is periodic with period N, and we get for its transform

$$(4\text{-}98)\qquad \mathscr{D}_N\{y_i\} = \sum_{i=0}^{N-1} y_i \, e^{-jik2\pi/N} = \sum_{i=0}^{M-1} x_i \, e^{-jik2\pi/N} =$$

$$= \sum_{i=0}^{M-1} x_i \, e^{-jik(M/N)\,2\pi/M} = X_{kM/N},$$

$$k = 0, 1, \dots, N-1.$$

The last expression represents formally a discrete Fourier transform of the original sequence. When using $kM/N = m$, a sequence X_k is defined for $m = 0, 1, \dots, M-1$ which determines values of Y_k for those indices $k = mN/M$ which are integers. From that results the following property.

Theorem 19: Relationship between the image of an original and a padded sequence

The elements of an image Y_k of a sequence y_i padded by relation (4-97), *the indices of which satisfy the condition $k = mN/M$ for $m = 0, 1, \dots, M-1$, are equal to corresponding terms of the image X_k of the original sequence x_i:*

$$(4\text{-}99)\qquad Y_{mN/M} = X_m, \qquad m = 0, 1, \dots, M-1.$$

From that, there further follows as a special case this theorem:

Theorem 20: Relationship between the image of an original sequence and a sequence padded to the p-multiple length

If we pad an original sequence of length M with a total of $(p-1)\,M$ zero points to the p-multiple length $N = pM$, the following then holds

$$(4\text{-}100)\qquad Y_{pm} = X_m, \qquad m = 0, 1, \dots, M-1,$$

and thence every p-th term of an image of the padded sequence is identical to an image of the original sequence.

The question arises as to what the remaining terms of the padded sequence are equal to; this will be shown in § 5.3.1.

We now show some properties of the image of a sequence padded to double length, i.e. for $p = 2$. In this case

$$(4\text{-}101)\qquad Y_{2m} = X_m, \qquad m = 0, 1, \dots, M-1,$$

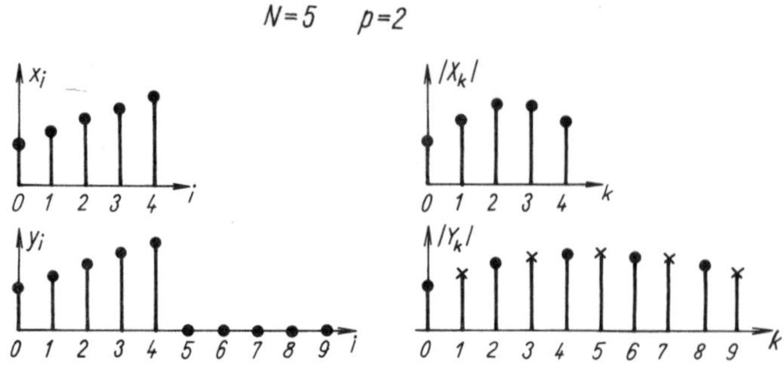

12. A sequence padded with zero terms to double length and its transform

which means the image of the original sequence is identical to even terms of the image of the padded sequence. In Figure 12 is shown the relation between the images of both sequences for $N = 5$ and $p = 2$. For odd terms we get in this case from definition (4-1) the expression

$$(4\text{-}102)\quad Y_{2m+1} = \sum_{i=0}^{M-1} x_i \,\mathrm{e}^{-\mathrm{j}i(2m+1)2\pi/(2M)} = \sum_{i=0}^{M-1} x_i \,\mathrm{e}^{-\mathrm{j}i\pi/M}\, \mathrm{e}^{\mathrm{j}im2\pi/M} = \mathscr{D}_M\{x_i \,\mathrm{e}^{-\mathrm{j}i\pi/M}\},$$

$$m = 0, 1, \dots, M-1,$$

from which it follows that the odd terms of the image are determined by the discrete Fourier transform of the sequence multiplied by an exponential expression that may be formally interpreted as a shift by a half interval.

Another possible interpretation follows from comparison with the so-called odd discrete Fourier transform introduced by J. L. Vernet [86].

$$(4\text{-}103)\quad Z_k = \sum_{i=0}^{M-1} x_i \,\mathrm{e}^{-\mathrm{j}i(2k+1)2\pi/(2M)}, \qquad k = 0, 1, \dots, M-1.$$

It appears from the comparison that

$$(4\text{-}104)\quad Y_{2m+1} = Z_m$$

so that odd terms of the image of a sequence $x_i \triangleq X_k$ padded with zero terms to double length are equal to the image of the sequence x_i in the sense of the odd discrete Fourier transform.

One can proceed similarly in the image and pad it with zero terms. An analogous process of padding with zero terms after the last term can be taken with complex sequences only.

If X_k is the image of a real sequence we must carry out the padding with zero terms so as not to violate this feature. This whole procedure will therefore be presented in a section on interpolation (§ 5.3.2).

4.3.2 Discrete Fourier transform of a repeated and a rarefied sequence

By *a repeated sequence* or more precisely a *p-times repeated sequence* x_i, $i = 0, 1, \ldots, N-1$ we shall understand a sequence y_i, $i = 0, 1, \ldots, pN-1$, set up by repeating the original sequence p-times

$$y_i = y_{\nu N+\mu} = x_\mu, \qquad \begin{matrix}\mu = 0, 1, \ldots, N-1, \\ \nu = 0, 1, \ldots, p-1.\end{matrix} \tag{4-105}$$

Although we consider the sequence x_i to be periodic, it is expedient, because the sequences specified are finite, to introduce the repeated sequence.

Furthermore, we introduce a *p-times rarefied sequence* $z_n \,\hat{=}\, Z_\varkappa$, $\varkappa = 0, 1, \ldots, L-1$, as a sequence that arises from a sequence $x_i \,\hat{=}\, X_k$, $i = 0, 1, \ldots, N-1$ by inserting after each term $p-1$ zero terms, so that

$$z_n = \begin{cases} x_i, & n = pi, \\ 0, & n \neq pi, \end{cases} \qquad i = 0, 1, \ldots, N-1. \tag{4-106}$$

We now derive some properties of the image of a p-times repeated sequence.

For the discrete Fourier transform of a repeated sequence we get

$$\begin{aligned} Y_\varkappa &= \sum_{i=0}^{pN-1} y_i \,\mathrm{e}^{-\mathrm{j}i\varkappa 2\pi/(pN)} = \\ &= \sum_{\nu=0}^{p-1} \sum_{i=\nu N}^{(\nu+1)N-1} y_i \,\mathrm{e}^{-\mathrm{j}i\varkappa 2\pi/(pN)} = \\ &= \sum_{\nu=0}^{p-1} \sum_{\mu=0}^{N-1} y_{\mu+\nu N} \,\mathrm{e}^{-\mathrm{j}(\mu+\nu N)2\pi/(pN)} = \\ &= \sum_{\mu=0}^{N-1} x_\mu \,\mathrm{e}^{-\mathrm{j}\mu\varkappa 2\pi/(pN)} \sum_{\nu=0}^{p-1} \mathrm{e}^{-\mathrm{j}\nu\varkappa 2\pi/p}. \end{aligned}$$

Since

$$\sum_{\nu=0}^{p-1} \mathrm{e}^{-\mathrm{j}\nu\varkappa 2\pi/p} = \begin{cases} p, & \varkappa = \lambda p, \\ 0, & \varkappa \neq \lambda p, \end{cases} \qquad \lambda = 0, \pm 1, \pm 2, \ldots,$$

we obtain the relation

$$Y_\varkappa = \begin{cases} pX_\lambda, & \varkappa = \lambda p, \\ 0, & \varkappa \neq \lambda p, \end{cases} \tag{4-107}$$

expressing the p-times rarefied sequence. From that, the following theorems result:

Theorem 21: Transform of a repeated sequence

The image of a sequence y_i set up by a p-times repeated sequence x_i, $i = 0, 1, \ldots, N-1$ is a sequence $Y_\varkappa$, which is equal to a p-times rarefied sequence pX_k.

Theorem 22: Inverse transform of a rarefied sequence

The original of a sequence $Y_\varkappa$, that is a p-times rarefied sequence X_k, is a sequence y_i set up by a p-times repeated sequence $(1/p)\,x_i$.

We also show some properties of the discrete Fourier transform of a rarefied sequence for which

$$(4\text{-}108)\quad Z_\varkappa = \sum_{n=0}^{L-1} z_n\, \mathrm{e}^{-\mathrm{j}n\varkappa 2\pi/L} = \sum_{i=0}^{N-1} z_{pi}\, \mathrm{e}^{-\mathrm{j}ip\varkappa 2\pi/L} = \sum_{i=0}^{N-1} x_i\, \mathrm{e}^{-\mathrm{j}i\varkappa 2\pi/N}, \qquad \varkappa = 0, 1, \ldots, L-1,$$

and we arrive at the conclusion

$$(4\text{-}109)\quad Z_\varkappa = X_\varkappa, \qquad \varkappa = 0, 1, \ldots, L-1.$$

This can be summed up in theorems 23 and 24.

Theorem 23: Transform of a rarefied sequence

A sequence z_n set up by a p-times rarefied sequence x_i has an image $Z_\varkappa$ which is the sequence equal to the p-times repeated image X_k of a sequence x_i.

Theorem 24: Inverse transform of a repeated sequence

The original z_n of a sequence $Z_\varkappa$ set up by a p-times repeated sequence X_k is a p-times rarefied sequence x_i.

A twice repeated sequence and its transform are shown in Figure 13.

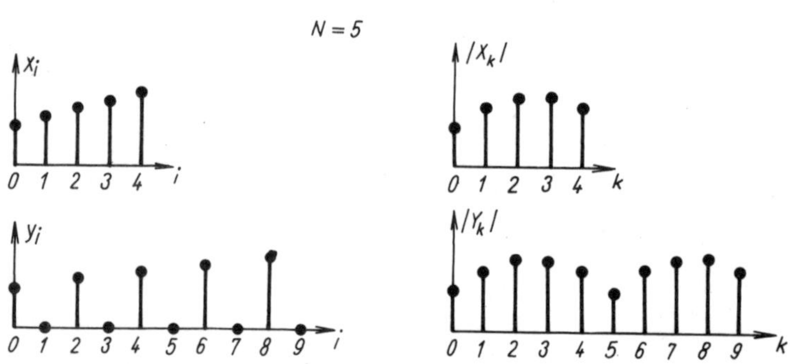

13. A twice rarefied sequence and its transform

Example 13. Let us consider a sequence $s_n \mathrel{\hat=} S_l$, $n = 0, 1, \ldots, N-1$, $N = pM$.

$$(4\text{-}110)\quad s_n = \begin{cases} 1, & n = \nu p, \\ 0, & n \neq \nu p, \end{cases} \qquad \nu = 0, 1, \ldots, M-1.$$

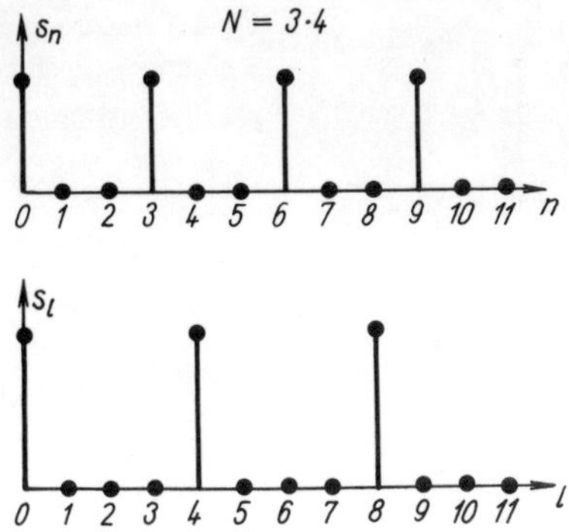

14. Relation between a rarefied and a repeated sequence and their transforms for $p = 3$, $M = 4$

If we take this sequence to be a p-times rarefied sequence $x_i = 1$, $i = 0, 1, \ldots, M-1$ with image $X_0 = M$, $X_k = 0$, $k = 1, 2, \ldots, M-1$, then its image is, according to theorem 23, a p-times repeated sequence X_k, i.e. the following applies:

$$(4\text{-}111)\quad S_l = \begin{cases} M, & l = \mu M, \\ 0, & l \neq \mu M, \end{cases} \qquad l = 0, 1, \ldots, p-1.$$

The sequence s_n may also, however, be taken to be an M-times repeated sequence $y_0 = 1, y_i = 0, i = 1, 2, \ldots, p-1$ with image $Y_k = 1, k = 0, 1, 2, \ldots, p-1$, so that the image S_l, by theorem 21, is equal to an M-times repeated sequence MY_k. The sequence s_n is shown together with its image in Figure 14.

4.3.3 Discrete Fourier transform of selected sequences

Let us have a sequence $x_i \triangleq X_k$, $i = 0, 1, \ldots, N-1$ and we suppose that $N = pM$ where pM are integer numbers. By a selected sequence we shall understand the sequence ${}_my_\nu$, $\nu = 0, 1, \ldots, M-1$, set up by every p-th term of a sequence x_i starting with the m-th. From the sequence x_i p selected sequences can be created to which the following expression applies

$$(4\text{-}112)\quad {}_my_\nu = x_{\nu p + m}, \qquad \nu = 0, 1, \ldots, M-1, \quad m = 0, 1, \ldots, p-1.$$

An example of selected sequences for $p = 2$ and $m = 0$ and $m = 1$ is shown in Figure 15.

To show the properties of the image of a selected sequence, we use an inverse transform to express a general term of the sequence x_i for $i = \nu p + m$:

$$(4\text{-}113)\quad x_{\nu p + m} = \frac{1}{N}\sum_{k=0}^{N-1} X_k\, e^{jk(\nu p + m)2\pi/N}.$$

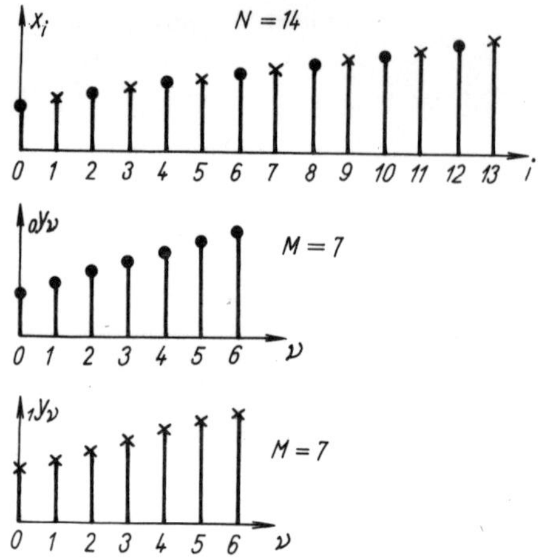

15. A selected sequence of even terms $_0y_n$ and odd terms $_1y_n$ of a sequence x_i

On using the periodicity of the expression $e^{jk2\pi/M}$ with period M we may write

$$x_{\nu p+m} = \frac{1}{N} \sum_{\mu=0}^{p-1} \sum_{p=\mu M}^{(\mu+1)M-1} X_k\, e^{jk\nu 2\pi/M}\, e^{jkm2\pi/N} =$$

$$= \frac{1}{N} \sum_{\mu=0}^{p-1} \sum_{\varkappa=0}^{M-1} X_{\varkappa+\mu M}\, e^{j(\varkappa+\mu M)(\nu p+m)\,2\pi/N}.$$

This expression can be adjusted to a form

$$x_{\nu p+m} = \frac{1}{M} \sum_{\varkappa=0}^{M-1} e^{j\varkappa m2\pi/N}\, \frac{1}{p} \sum_{\mu=0}^{p-1} X_{\varkappa+\mu M}\, e^{j\mu m2\pi/p}\, e^{j\nu\varkappa 2\pi/M}, \tag{4-114}$$

from which it follows that

$$x_{\nu p+m} = \mathscr{D}_M^{-1} \left\{ e^{j\varkappa m2\pi/N}\, \frac{1}{p} \sum_{\mu=0}^{p-1} X_{\varkappa+\mu M}\, e^{j\mu m2\pi/p} \right\}. \tag{4-115}$$

After introducing the notation $\mathscr{D}_M\{x_{\nu p+m}\} = {}_mY_\varkappa$ for the image of an m-th selected sequence ${}_my_\nu x_{\nu p+m}$, we get for the image of a selected sequence the expression

$${}_mY_\varkappa = e^{j\varkappa m2\pi/N}\, \frac{1}{p} \sum_{\mu=0}^{p-1} X_{\varkappa+\mu M}\, e^{j\mu m2\pi/p}, \qquad \varkappa = 0, 1, \dots, M-1. \tag{4-116}$$

Theorem 25: Transform of a selected sequence

Every $\varkappa$-th element of an image sequence ${}_mY_\varkappa$ of an m-th selected sequence ${}_my_\nu = x_{\nu p+m}$ is equal to an $e^{j\varkappa m2\pi/N}$-multiple of the inverse transform of a selected sequence $X_{\varkappa+\mu M}$ set up by every $M = (N/p)$-th term, starting with the $\varkappa$-th one:

$${}_mY_\varkappa = e^{j\varkappa m2\pi/N}\, \mathscr{D}_p^{-1}\{X_{\varkappa+\mu M}\}, \qquad \varkappa = 0, 1, \dots, M-1, \tag{4-117}$$

We now present as an example an expression for the image of a sequence of even terms and a sequence of odd terms. These sequences correspond to selected sequences for $p = 2$ and $m = 0$, 1 presented in Figure 15. For the image of the selected sequence of even terms we get for $\varkappa = 0, 1, \dots, N/2 - 1$

(4-118) $\quad {}_0Y_\varkappa = \mathscr{D}_2^{-1}\{X_{\varkappa+\mu N/2}\} = \frac{1}{2}(X_\varkappa + X_{\varkappa+N/2})$

and for the image of the selected odd terms the sequence

(4-119) $\quad {}_1Y_\varkappa = \mathscr{D}_2^{-1}\{X_{\varkappa+\mu N/2}\, \mathrm{e}^{\mathrm{j}\varkappa 2\pi/N}\} = \frac{1}{2}(X_\varkappa - X_{\varkappa+N/2})\, \mathrm{e}^{\mathrm{j}\varkappa 2\pi/N}.$

4.3.4 Expressing the discrete Fourier transform of a sequence by means of a discrete Fourier transform of selected sequences

We begin with a sequence $x_i \triangleq X_k$, $i = 0, 1, \dots, N - 1$, $N = pM$, which we split into selected sequences set up so that from sequence x_i we take successively every p-th term starting with the m-th. We therefore construct p selected sequences ${}_mx_i$, $m = 0, 1, \dots, p - 1$, of length $M = N/p$. On this basis we can express the discrete Fourier transform of the sequence x_i in the following manner:

(4-120)
$$\begin{aligned} X_k &= \sum_{m=0}^{p-1} \sum_{i=m+\nu p}^{m+(M-1)p} x_i\, \mathrm{e}^{-\mathrm{j}ki2\pi/N} = \\ &= \sum_{m=0}^{p-1} \sum_{\nu=0}^{M-1} x_{m+\nu p}\, \mathrm{e}^{-\mathrm{j}k(m+\nu p)\,2\pi/N} = \\ &= \sum_{m=0}^{p-1} \sum_{\nu=0}^{M-1} x_{\nu p+m}\, \mathrm{e}^{-\mathrm{j}k\nu 2\pi/M}\, \mathrm{e}^{-\mathrm{j}km2\pi/N}, \\ &\qquad k = 0, 1, \dots, N - 1. \end{aligned}$$

On expressing the index k in the form $k = \mu M + n$, $\mu = 0, 1, \dots, p - 1$, $n = 0, 1, \dots, M - 1$, we get the important expression

(4-121) $$X_{\mu M+n} = \sum_{m=0}^{p-1} [\mathrm{e}^{-\mathrm{j}mn2\pi/N}(\sum_{\nu=0}^{M-1} x_{\nu p+m}\, \mathrm{e}^{-\mathrm{j}\nu n2\pi/M})]\, \mathrm{e}^{-\mathrm{j}\mu m2\pi/p}.$$

We can imagine the splitting of the sequence x_i into the selected sequences in matrix form, when particular selected sequences set up rows of a matrix of type (p, M). We write general a term of the matrix as $\hat{x}_{m,\nu} = x_{\nu p+m}$.

By analogy, an image sequence can also be split into M selected sequences of length p. We could therefore write it again as the matrix whose rows are set up by altogether M selected sequences, i.e. as a matrix of type (M, p). For a better illustration we use instead of this matrix the transposed matrix in which selected sequences appear as columns, that is a matrix of type (p, M).

Analysing now expression (4-121) it is obvious that it can be divided into the following operations:

1. A discrete Fourier transform of the selected sequences setting up single rows:

$$(4\text{-}122)\quad \sum_{\nu=0}^{M-1} x_{\nu p+m}\, \mathrm{e}^{-\mathrm{j}\nu n 2\pi/M} = \mathscr{D}_M\{x_{\nu p+m}\} = X_{m,n}, \qquad m = 0, 1, \dots, p-1, \quad n = 0, 1, \dots, M-1.$$

The discrete Fourier transform of each row can again be imagined as a row of a new matrix of type (p, M) with general term $X_{m,n}$.

2. Multiplying every term $X_{m,n}$ of this matrix by the term $\mathrm{e}^{-\mathrm{j}mn2\pi/N}$ and so creating a matrix with general term

$$(4\text{-}123)\quad Z_{m,n} = X_{m,n}\, \mathrm{e}^{-\mathrm{j}mn2\pi/N}.$$

3. Discrete Fourier transforms of sequences which set up single columns

$$(4\text{-}124)\quad \sum_{m=0}^{p-1} Z_{m,n}\, \mathrm{e}^{-\mathrm{j}m\mu 2\pi/p} = \mathscr{D}_p\{Z_{m,n}\} = X_{\mu M+n}, \qquad \mu = 0, 1, \dots, p-1, \quad n = 0, 1, \dots, M-1.$$

The discrete Fourier transform of the sequence x_i is then determined by a sequence set up by the sequentially ordered rows of the matrix with general term $Z_{m,n}$.

We can also arrive at expression (4-122), created by a purely formal procedure, by splitting the sequence x_i into selected sequences. We denote the image of an m-th selected sequence ${}_m x_{\nu p}$ as ${}_m X_k$, $m = 0, 1, \dots, p-1$, $k = 0, 1, \dots, M-1$. Every m-th selected sequence we now rarefy p-times. Then for its image

$$(4\text{-}125)\quad {}_m x_{\nu p} \mathrel{\hat{=}} {}_m X_k, \qquad m = 0, 1, \dots, p-1, \quad k = 0, 1, \dots, N-1.$$

Every m-th sequence so created we shift right by m positions so that it starts with the first nonzero term in the m-th position. For the image of that sequence we then get ${}_m X_k\, \mathrm{e}^{-\mathrm{j}mk2\pi/N}$ by theorem 11 on the shift. The sequence x_i is obtained by a superposition of all the rarefied selected sequences shifted by m terms and for its image

$$(4\text{-}126)\quad X_k = \sum_{m=0}^{p-1} {}_m X_k\, \mathrm{e}^{-\mathrm{j}mk2\pi/N} = \sum_{m=0}^{p-1} \Big(\sum_{\nu=0}^{M-1} x_{\nu p+m}\, \mathrm{e}^{-\mathrm{j}\nu k 2\pi/M} \Big)\, \mathrm{e}^{-\mathrm{j}mk2\pi/N}.$$

Thus we can state that the discrete Fourier transform of a sequence of length $N = pM$ can be computed by carrying out p-times a transform of selected sequences of length M and M-times a transform of a sequence of length p. As we shall see later in Chapter 5, this split corresponds closely to the two-dimensional discrete Fourier transform and is most important in providing a basis for reducing the number of computational steps and so creating the algorithm of discrete Fourier transform computation known as the fast Fourier transform.

4.3.5 Discrete Fourier transform of a cyclically permuted sequence

By a cyclically permuted sequence [45] $y_n \triangleq Y_m$, $n = 0, 1, \ldots, N-1$ to a sequence $x_i \triangleq X_k$, $i = 0, 1, \ldots, N-1$, we shall understand the sequence that arises by a cyclic selection of every p-th term in the case where the numbers p and N are mutually prime. For y_n it therefore follows that

$$(4\text{-}127) \quad y_n = x_{(np)_N}, \qquad n = 0, 1, \ldots, N-1,$$

where the symbol $(np)_N$ means the remainder after division of the number np by the number N:

$$(4\text{-}128) \quad i = np \bmod N.$$

So for the example of $N = 9$ and $p = 4$, when we select cyclically from x_i every fourth term, the sequence y_n is expressed by means of x_i by the following table

y_n	y_0	y_1	y_2	y_3	y_4	y_5	y_6	y_7	y_8
$x_{(4n)_9}$	x_0	x_4	x_8	x_3	x_7	x_2	x_6	x_1	x_5.

For the transform of sequence y_n holds

$$(4\text{-}129) \quad Y_m = \sum_{n=0}^{N-1} y_n \, e^{-jnm2\pi/N} = \sum_{n=0}^{N-1} x_{(np)_N} \, e^{-jnm2\pi/N}.$$

Let us now use q for the number inverse mod N to the number p, i.e. the number to which the following applies

$$(4\text{-}130) \quad (pq)_N = 1.$$

Calculation of the number q can be carried out by means of the solution of congruences [61]. One of these methods based upon Euler's function leads in our case to the relation

$$q = (p^{\varphi(N)-1})_N$$

where $\varphi(N)$ is Euler's function

$$\varphi(N) = N \prod_{\nu=1}^{k} (1 - r_\nu^{-1})$$

if the canonical expansion $N = \prod_{\nu=1}^{k} r_\nu^{\varkappa_\nu}$ holds for N.

Thus, for example, for $N = 9$ the canonical expansion $N = 3^2$ holds, i.e. $r_1 = 3$, $\varkappa_1 = 2$, and for Euler's function we get $\varphi(9) = 9(1 - 1/3) = 6$. From that follows $q = (4^5)_9 = 7$.

In our case $q = 7$. In relation (4-129) we now substitute $n = (ql)_N$. In that case a sequence of numbers $l = 0, 1, \ldots, N-1$ changes to a sequence of the same numbers but in a different order. As a result of this substitution we get

$$(4\text{-}131)\quad Y_m = \sum_{l=0}^{N-1} x_{((ql)_N p)_N}\, \mathrm{e}^{-\mathrm{j}m(ql)_N 2\pi/N}.$$

Since the following relation holds,

$$(4\text{-}132)\quad (ab)_N = ((a)_N b)_N ,\dagger)$$

we can modify an index at x in relation (4-131)

$$(4\text{-}133)\quad (np)_N = ((ql)_N p)_N = (lqp)_N = (l(qp)_N)_N = (l)_N$$

and furthermore an expression in the exponent

$$(4\text{-}134)\quad m(ql)_N = (m(ql)_N)_N = (mql)_N = ((mq)_N l)_N = (mq)_N l.$$

From relation (4-131) for Y_m we then get

$$(4\text{-}135)\quad Y_m = \sum_{l=0}^{N-1} x_{(l)_N}\, \mathrm{e}^{-\mathrm{j}l(mq)_N 2\pi/N} = $$

$$= \sum_{l=0}^{N-1} x_l\, \mathrm{e}^{-\mathrm{j}l(mq)_N 2\pi/N},$$

and the image of the cyclically permuted sequence thus obeys

$$(4\text{-}136)\quad Y_m = X_{(mq)_N}.$$

In our case $N = 9$, $p = 4$, $q = 7$ and thus Y_m is expressed from X_k by this table:

Y_m	Y_0	Y_1	Y_2	Y_3	Y_4	Y_5	Y_6	Y_7	Y_8
$X_{(7m)_9}$	X_0	X_7	X_5	X_3	X_1	X_8	X_6	X_4	X_2.

When choosing $p = N - 1$, then also $q = N - 1$. Because $(ip)_N = (i(N-1))_N = -i$, then $y_n = x_{-n}$. As $(mq)_N = (n(N-1))_N = -m$, then $Y_m = X_{-m}$ and this is in agreement with relation (4-17) which we have derived for the image of a reversed sequence.

---

†) Let us put $a = a_1 N + a_2$; then $(a)_N = a_2$. For a product $(ab)_N$ there holds $(ab)_N = ((a_1 N + a_2)\, b)_N = (a_1 bN + a_2 b)_N = (a_2 b)_N = ((a)_N\, b)_N$.

5. OTHER PROPERTIES OF DISCRETE FOURIER TRANSFORMS AND THEIR USE IN COMPUTING FOURIER TRANSFORMS

5.1 Relationship between discrete Fourier transforms, Fourier transforms and Fourier series

5.1.1 Relationship between discrete Fourier transforms and Fourier transforms

Let us assume that x_p is a periodic function with period P and Fourier series

$$x_p(t) = \sum_{\varkappa=-\infty}^{\infty} C_\varkappa \, \mathrm{e}^{\mathrm{j}\varkappa 2\pi t/P}. \tag{5-1}$$

Using $x_i = x(iT)$, $T = P/N$, $i = 0, 1, \ldots, N-1$, expressing x_i from (5-1) and substituting into the defining relation for the discrete Fourier transform, we obtain

$$X_k = \sum_{i=0}^{N-1} x_i \, \mathrm{e}^{-\mathrm{j}ik2\pi/N} = \sum_{\varkappa=-\infty}^{\infty} C_\varkappa \sum_{i=0}^{N-1} \mathrm{e}^{\mathrm{j}i(\varkappa-k)\,2\pi/N}.$$

Since

$$\sum_{i=0}^{N-1} \mathrm{e}^{\mathrm{j}i(\varkappa-k)\,2\pi/N} = \begin{cases} N, & \varkappa - k = \nu N, \\ 0, & \varkappa - k \neq \nu N, \end{cases} \quad \nu = 0, \pm 1, \pm 2, \ldots,$$

we obtain an expression for X_k

$$X_k = N \sum_{\nu=-\infty}^{\infty} C_{k+\nu N}. \tag{5-2}$$ †)

On using x_0 for the function x_p on a basic period,

$$x_0(t) = \begin{cases} x_p(t), & t \in [0, P), \\ 0, & t \notin [0, P), \end{cases}$$

and $X_0(\omega) \mathrel{\hat{=}} x_0(t)$, then remembering from (2-64) that

$$C_k = \frac{1}{NT} X_0\left(\frac{k \,.\, 2\pi}{NT}\right),$$

---

†) One can also arrive at this expression from (2-32) by considering the relationship between D_k and X_k described in § 5.1.3.

we may also express (5-2) as

$$(5\text{-}3) \qquad X_k = \frac{1}{T}\sum_{\nu=-\infty}^{\infty} X_0\left((k+\nu N)\frac{2\pi}{NT}\right).$$

Relations (5-2) and (5-3) indicate that a direct expression of the function X_0 and coefficients C_k by means of coefficients X_k is not generally possible. Furthermore, from these relations one can see that the approximate expressions $C_k \approx N^{-1}X_k$ and $X_0(k\,.\,2\pi/(NT)) \approx TX_k$ depend on the rate of decrease with increasing k of the coefficients C_k, or how the function X_0 decreases with increasing ω. It is obvious that only in the case where

$$(5\text{-}4) \qquad X_0(\omega) = 0, \qquad |\omega| > \pi/T,$$

or

$$(5\text{-}5) \qquad C_k = 0, \qquad |k| > N/2,$$

is the following also true

$$(5\text{-}6) \qquad X_0\left(\frac{k\,.\,2\pi}{NT}\right) = TX_k$$

or

$$(5\text{-}7) \qquad C_k = N^{-1}X_k.$$

Example 14. A sequence x_i of function values $x(iT)$, $i = 0, 1, \ldots, N-1$, $T > 0$, of a function $x(t) = \mathrm{e}^{-at}$, $a \neq 0$ possesses a transform that we obtain by using (4-1)

$$X_k = \sum_{i=0}^{N-1} \mathrm{e}^{-aiT}\,\mathrm{e}^{-\mathrm{j}ik2\pi/N} = \sum_{i=0}^{N-1} \mathrm{e}^{-(aT+\mathrm{j}k2\pi/N)i} =$$

$$= \frac{1-\mathrm{e}^{-aTN}}{1-\mathrm{e}^{-(aT+\mathrm{j}k2\pi/N)}}, \qquad k = 0, 1, \ldots, N-1.$$

Therefore

$$\operatorname{Re} X_k = (1-\mathrm{e}^{-aTN})\frac{1-\mathrm{e}^{-aT}\cos(k\,.\,2\pi/N)}{1+\mathrm{e}^{-2aT}-2\mathrm{e}^{-aT}\cos(k\,.\,2\pi/N)},$$

$$\operatorname{Im} X_k = (1-\mathrm{e}^{-aTN})\frac{\mathrm{e}^{-aT}\sin(k\,.\,2\pi/N)}{1+\mathrm{e}^{-2aT}-2\mathrm{e}^{-aT}\cos(k\,.\,2\pi/N)}.$$

For comparison we present the Fourier transform $X_0(\omega)$ of the finite function $x_0(t) = \mathrm{e}^{-at}$, $0 < t < P = NT$, $x_0(t) = 0$, $t < 0$, $t > P$:

$$\operatorname{Re} X_0(\omega) = \frac{-\mathrm{e}^{-aP}(a\cos\omega P - \omega\sin\omega P) + a}{a^2+\omega^2},$$

$$\operatorname{Im} X_0(\omega) = \frac{\mathrm{e}^{-aP}(\omega\cos\omega P + a\sin\omega P) - \omega}{a^2+\omega^2},$$

and the Fourier transform $X_1(\omega)$ of the function $x_1(t) = e^{-at}$, $t \geqq 0$, $x_1(t) = 0$, $t < 0$:

$$\operatorname{Re} X_1(\omega) = a(a^2 + \omega^2)^{-1},$$
$$\operatorname{Im} X_1(\omega) = -\omega(a^2 + \omega^2)^{-1}.$$

We show that relation (5-2) can be efficiently used in some cases to obtain an expression for the discrete Fourier transform.

Example 15. In example 2, we have obtained for the coefficients of the Fourier series the expression $C_k = -\mathrm{j} \,.\, 16/(\pi k)^3$ for odd k and $C_k = 0$ for $k = 0$ and k even. If we construct from the function f of example 2 the sequence $x_i = f(iT)$, $T = P/N$, $i = 0, 1, \ldots, N-1$, then from relation (5-2) we get for the discrete Fourier transform of this sequence

$$X_k = N \sum_{\nu=-\infty}^{\infty} C_{k+\nu N} = -\mathrm{j}\frac{16N}{\pi^3} \sum_{\nu=-\infty}^{\infty} \frac{1}{(k+\nu N)^3}.$$

Since, from (5-46)

$$\sum_{\nu=-\infty}^{\infty} \frac{1}{(x-\nu u)^3} = \left(\frac{\pi/u}{\sin \pi x/u}\right)^3 \cos \pi x/u,$$

we obtain for odd k

$$X_k = -\mathrm{j}\frac{16N}{\pi^3}\frac{1}{k^3}\left(\frac{k\,.\,\pi/N}{\sin(k\,.\,\pi/N)}\right)^3 \cos(k\,.\,\pi/N)$$

and $X_k = 0$ for even k.

Example 16. Beginning with the function

$$x(t) = \begin{cases} 1, & t_1 < t < t_2, \\ 0, & t < t_1,\ t_2 < t < P, \\ \frac{1}{2}, & t = t_1,\ t = t_2, \end{cases}$$

where $t_1 = i_1 T$ and $t_2 = i_2 T$, $0 < i_1 < i_2 < N$, $P = NT$, we obtain for the coefficients of its Fourier series

$$C_k = (k\,.\,\pi)^{-1} \mathrm{e}^{-\mathrm{j}k(\alpha_1+\alpha_2)\pi} \sin k(\alpha_2 - \alpha_1)\,\pi, \qquad k = \pm 1, \pm 2, \ldots,$$

$$C_0 = \alpha_2 - \alpha_1,$$

where $\alpha_1 = i_1/N$ and $\alpha_2 = i_2/N$.

For the discrete Fourier transform of the sequence $x_i = x(iT)$ we get by using (5-2)

$$X_k = \mathrm{e}^{-\mathrm{j}k(i_2+i_1)\pi/N} \sin k(i_2 - i_1)\,\pi/N \operatorname{cotg} k\,.\,\pi/N, \qquad k = 0, 1, \ldots, N-1.$$

5.1.2 Discrete Fourier transforms and a numerical computation of Fourier transforms

If we express the Fourier transform of a finite function $x(t) \triangleq X(\omega)$ from its function values $x(iT) \triangleq x_i$, $i = 0, 1, \ldots, N$, then the trapezoid rule gives

$$(5\text{-}8) \qquad {}_LX(\omega) = T[\sum_{i=0}^{N-1} x_i \, \mathrm{e}^{-\mathrm{j}\omega iT} + \tfrac{1}{2}(x_N - x_0)].$$

If we substitute $\omega = k\,.\,2\pi/(NT)$, we get

$$(5\text{-}9) \qquad {}_LX\left(\frac{k\,.\,2\pi}{NT}\right) = TX_k + \frac{(x_N - x_0)\,T}{2}.$$

By using a trapezoid rule with inscribed rectangles we arrive at the expression

$$(5\text{-}10) \qquad {}_0X\left(\frac{k\,.\,2\pi}{NT}\right) = TX_k.$$

Because of what was said in Chapter 3, it follows from (5-10) that the discrete Fourier transform is only a very loose approximation to the function X. A better approximation, as a rule, requires the use of Richardson's extrapolation [67], [34], [54].

For the application of Richardson's extrapolation, it is necessary to carry out a computation of the discrete Fourier transform of two sequences of different lengths $N_1 > N_2$ set up by function values of the function x within the same interval. Using

$$ {}_1x_i \triangleq {}_1X_k, \qquad {}_1x_i = x(iT_1), \qquad T_1 = P/N_1, \qquad i = 0, 1, \ldots, N_1 - 1,$$

$$ {}_2x_i \triangleq {}_2X_k, \qquad {}_2x_i = x(iT_2), \qquad T_2 = P/N_2, \qquad i = 0, 1, \ldots, N_2 - 1,$$

one can determine extrapolated values ${}_RX(\omega)$ for $\omega = \omega_k = k\,.\,2\pi/P$ by the relation

$$ {}_RX(\omega_k) = \alpha_1 X_k + (1 - \alpha)_2 X_k, \qquad k = 0, 1, \ldots, N - 1,$$

where $\alpha = N_1^2/(N_1^2 - N_2^2)$.

The computation can be simplified further by choosing for the sequence ${}_2x_l$ the subsequence of even terms of the sequence ${}_1x_i$, i.e. the sequence ${}_1x_0, {}_1x_2, \ldots, {}_1x_{N-2}$.

This manner of extrapolation eliminates a computational error of the order of N^{-2}. If we determine in addition a discrete Fourier transform ${}_3X_k$ of a selected subsequence of e.g. even terms from the subsequence of even terms, then the extrapolating relation

$$_RX(\omega_k) = \frac{64}{45}\,{}_1X_k - \frac{20}{45}\,{}_2X_k + \frac{1}{45}\,{}_3X_k, \qquad k = 0, 1, \ldots, N-1.$$

removes an error of the order of N^{-2} and of N^{-4}.

5.1.3 Discrete Fourier transform and an approximation by a trigonometric polynomial

In § 2.2.2 expressions for coefficients D_k have been presented which guarantee that a trigonometric polynomial

$$(5\text{-}11)\qquad u(t) = \sum_{k=-M}^{M} D_k\, e^{jk2\pi t/P}$$

satisfies at points $t = iT$, $T = P/N$, $i = 0, 1, \ldots, N-1$ the condition of a minimal sum of squares of absolute values of deviations $x_i - u(iT)$. If $N = 2M + 1$, then for coefficients D_k we get

$$(5\text{-}12)\qquad D_k = N^{-1}\sum_{i=0}^{N-1} x_i\, e^{-jik2\pi/N}, \qquad k = 0, \pm 1, \ldots, \pm L,$$

for $L = (N-1)/2$. If $N = 2M$, then (5-11) holds for $L = N/2 - 1$ and for $k = N/2$

$$(5\text{-}13)\qquad D_{N/2} = D_{-N/2} = (2N)^{-1}\sum_{i=0}^{N-1} x_i(-1)^i.$$

Since (5-12) expresses the discrete Fourier transform, D_k may be expressed by means of the discrete Fourier transform of the sequence x_i:

$$(5\text{-}14a)\qquad D_k = \begin{cases} N^{-1}X_k, & k = 0, 1, \ldots, N/2 - 1, \\ (2N)^{-1}X_k, & k = \pm N/2, \\ N^{-1}X_{N+k}, & k = -1, -2, \ldots, -(N/2 - 1), \end{cases} \qquad N = 2M,$$

$$(5\text{-}14b)\qquad D_k = \begin{cases} N^{-1}X_k, & k = 0, 1, \ldots, (N-1)/2, \\ N^{-1}X_{N+k}, & k = -1, -2, \ldots, -(N-1)/2, \end{cases} \qquad N = 2M + 1.$$

Conversely, x_k can be expressed by means of D_k:

$$(5\text{-}15a)\qquad X_k = \begin{cases} ND_k, & k = 0, 1, \ldots, N/2 - 1, \\ ND_{k-N}, & k = N/2 + 1, \ldots, N-1, \\ 2ND_{N/2}, & k = N/2, \end{cases} \qquad M = N/2,$$

$$(5\text{-}15b)\qquad X_k = \begin{cases} ND_k, & k = 0, 1, \ldots, (N-1)/2, \\ ND_{k-N}, & k = (N+1)/2, \ldots, N-1, \end{cases} \qquad M = (N-1)/2.$$

5.1.4 Relationship of the discrete Fourier transform to the $\mathscr{Z}$ transform

The $\mathscr{Z}$ transform of a sequence x_i is defined [87], [66] by the relation

$$X_z(z) = \sum_{i=-\infty}^{\infty} x_i z^{-i}.$$

For a finite nonperiodic sequence we then get

$$X_z(z) = \sum_{i=0}^{N-1} x_i z^{-i}. \tag{5-16}$$

It follows that the discrete Fourier transform X_k is equal to the $\mathscr{Z}$ transform evaluated at points $z = \mathrm{e}^{\mathrm{j}k2\pi/N}$ on the unit circle:

$$X_k = X_z(\mathrm{e}^{-\mathrm{j}k2\pi/N}), \qquad k = 0, 1, \ldots, N-1. \tag{5-17}$$

After substituting for x_i into (5-16) from the relation for the inverse Fourier transform, we arrive at an expression for the $\mathscr{Z}$ transform in terms of the discrete Fourier transform:

$$X_z(z) = \frac{1 - Z^{-N}}{N} \sum_{k=0}^{N-1} X_k \frac{1}{1 - \mathrm{e}^{\mathrm{j}k2\pi/N} z^{-1}} \tag{5-18}$$

for the finite sequence x_i.

5.2 Relationship of the discrete Fourier transform to a Fourier transform of the symbolic expression of function values at single points

5.2.1 The Fourier transform of the symbolic expression of function values at single points

If we determine from a function x its function values $x(iT) = x_i$, $i \in \in (-\infty, \infty)$, we obtain a sequence x_i. This procedure can be given a physical interpretation known as the sampling of signals; see e.g. [66], [67], [68]. Mathematically this can be expressed as multiplication of a function x by a function r_n,

$$r_n(t) = \sum_{i=-\infty}^{\infty} q_n(t - iT), \tag{5-19}$$

where q_n is the δ sequence from § 2.3.1 denoted 1. Thus we create a signal

$$\tilde{x}_n(t) = x(t)\, r_n(t), \tag{5-20}$$

which is usually called *the sampled signal.* If we now "narrow" the function q_n so that we let n grow beyond all limits, then with respect to (2-33) there holds

$$\text{(5-21)} \quad \lim_{n\to\infty} r_n(t) = \lim_{n\to\infty} \sum_{i=-\infty}^{\infty} q_n(t - iT) = \sum_{i=-\infty}^{\infty} \delta(t - iT),$$

and for (5-20) we get

$$\text{(5-22)} \quad \lim_{n\to\infty} \tilde{x}_n(t) = \tilde{x}(t) = x(t) \sum_{i=-\infty}^{\infty} \delta(t - iT).$$

The signal $\tilde{x}$ is called *an ideally sampled (discretised) signal.* On using the property (2-35) of the product of a function with the δ distribution, (5-22) changes to an equivalent form

$$\text{(5-23)} \quad \tilde{x}(t) = \sum_{i=-\infty}^{\infty} x(iT)\, \delta(t - iT).$$

Expressions (5-22) and (5-23) can also be considered to be symbolic expressions of sequences $x(iT)$.

Since we have introduced in § 2.4.4 the Fourier transform of the δ distribution and of the periodic δ distribution, we can determine also a Fourier transform of the expression for the function $\tilde{x}$. Denoting $\tilde{x}(t) \triangleq \tilde{X}(\omega)$, then from (5-23) using (2-53) we get

$$\text{(5-24a)} \quad \tilde{X}(\omega) = \sum_{i=-\infty}^{\infty} x(iT)\, \mathrm{e}^{-\mathrm{j}\omega iT}.$$

If we use the theorem on the Fourier transform of a product (2-49), relation (2-56) and identity (2-59), we arrive from (5-22) at the expression

$$\tilde{X}(\omega) = \frac{1}{2\pi} X(\omega) * \frac{2\pi}{T} \sum_{\varkappa=-\infty}^{\infty} \delta\left(\omega - \varkappa \frac{2\pi}{T}\right).$$

The convolution properties (2-37) of the δ distribution then permit us to obtain from the latter relation an alternative expression for the Fourier transform in the form

$$\text{(5-24b)} \quad \tilde{X}(\omega) = \frac{1}{T} \sum_{\varkappa=-\infty}^{\infty} X\left(\omega - \varkappa \frac{2\pi}{T}\right).$$

5.2.2 Derivation of the discrete Fourier transform

1. Let us assume that x is a finite function and $x(iT) = x_i$, $i = 0, 1, \ldots, N - 1$. On using the symbolic expression (5-23)

$$\text{(5-25)} \quad \tilde{x}(t) = \sum_{i=0}^{N-1} x(iT)\, \delta(t - iT),$$

we obtain its Fourier transform according to (5-24a)

$$\text{(5-26)} \quad \tilde{X}(\omega) = \sum_{i=0}^{N-1} x(iT)\, \mathrm{e}^{-\mathrm{j}\omega iT}.$$

On replacing $\omega = k \,.\, 2\pi/(NT)$, we arrive at the conclusion that *the discrete Fourier transform of a finite sequence $x_i = x(iT)$ is equal to the Fourier transform of its symbolic expression evaluated at $\omega = k \,.\, 2\pi/(NT)$*:

$$(5\text{-}27) \qquad X_k = \tilde{X}\left(k \frac{2\pi}{NT}\right).$$

When replacing $\tilde{X}(\omega)$ from relation (5-24b), we can write (5-27) as

$$(5\text{-}28) \qquad X_k = T^{-1} \sum_{\varkappa=-\infty}^{\infty} X\left((k - \varkappa N)\frac{2\pi}{NT}\right).$$

2. Let us now assume that $x(t) \mathrel{\hat{=}} X(\omega)$ is not finite, but is an infinite sum of shifted functions $x(t - \nu P)$ which create a function x_p, by analogy with what we did in § 2.4.4:

$$x_p(t) = \sum_{\mu=-\infty}^{\infty} x(t - \mu P).$$

Now, when specifying values of this function for iT, $T = P/N$, $i = 0, 1, \ldots, N - 1$, we get

$$(5\text{-}29) \qquad x_p(iT) = \sum_{\mu=-\infty}^{\infty} x((i + \mu N)\, T).$$

For the Fourier transform of a function $\tilde{x}$ according to (5-24a) the following holds:

$$\tilde{X}(\omega) = \sum_{\nu=-\infty}^{\infty} x(\nu T)\, \mathrm{e}^{-\mathrm{j}\omega\nu T}.$$

Writing out the infinite sum into finite sums of single period N of the values $x(T)$, we get the relation

$$\tilde{X}(\omega) = \sum_{\mu=-\infty}^{\infty} \sum_{\nu=\mu N}^{(\mu+1)N-1} x(\nu T)\, \mathrm{e}^{-\mathrm{j}\omega\nu T},$$

which by the substitution $i = \nu - \mu N$ turns into the relation

$$\tilde{X}(\omega) = \sum_{i=0}^{N-1} \sum_{\mu=-\infty}^{\infty} x((i + \mu N)\, T)\, \mathrm{e}^{-\mathrm{j}\omega iT}\, \mathrm{e}^{-\mathrm{j}\omega\mu NT}.$$

Now using in addition (5-29), we obtain

$$\tilde{X}(\omega) = \sum_{i=0}^{N-1} x_p(iT)\, \mathrm{e}^{-\mathrm{j}\omega iT}\, \mathrm{e}^{-\mathrm{j}\omega\mu NT}.$$

For $\omega = k \,.\, 2\pi/(NT)$ we get

$$(5\text{-}30) \qquad \tilde{X}\left(k \frac{2\pi}{NT}\right) = \sum_{i=0}^{N-1} x_p(iT)\, \mathrm{e}^{-\mathrm{j}ik\, 2\pi/N},$$

which is an expression for the direct discrete Fourier transform. This relation is illustrated in Figure 16. It thus follows that *the discrete Fourier transform*

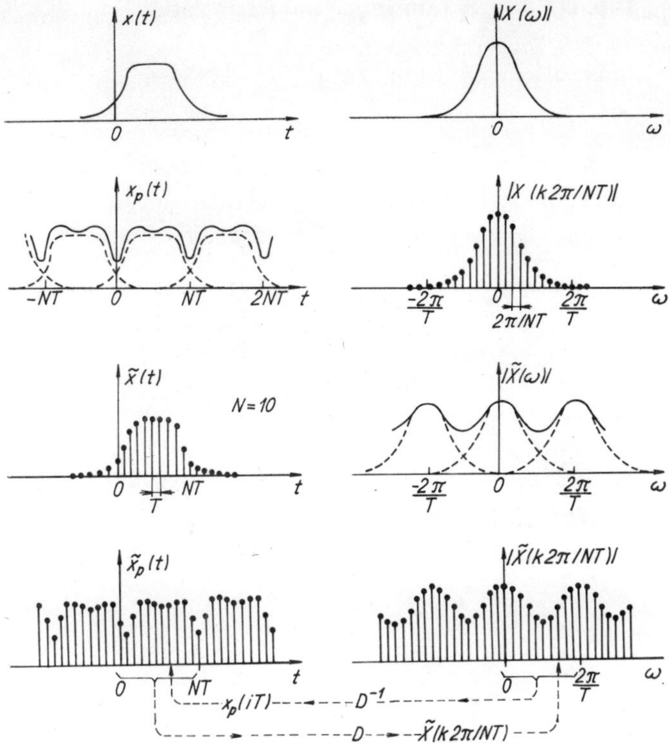

16. Interpretation of the discrete Fourier transform for an even function x

of the sequence of function values $x_p(iT)$, $i = 0, 1, \ldots, N-1$, *of a function* x_p *is equal to the Fourier transform of the symbolic expression of function values* $x(iT)$, $i \in (-\infty, \infty)$ *of the function* x *evaluated at* $\omega = k\,.\,2\pi/(NT)$.

3. In a similar way one can also arrive at the inverse discrete Fourier transform. We begin again from the function x_p, which we express by means of (2-64):

$$x_p(t) = P^{-1} \sum_{k=-\infty}^{\infty} X\left(k\frac{2\pi}{P}\right) e^{jk2\pi t/P}.$$

On writing out the sum into sections of N terms for $P = NT$ we get for $t = iT$

$$x_p(iT) = (NT)^{-1} \sum_{k=0}^{N-1} \sum_{\nu=-\infty}^{\infty} X\left((k+\nu N)\frac{2\pi}{NT}\right) e^{-jk2\pi/N}.$$

By using (5-24b) we get for $\omega = k\,.\,2\pi/(NT)$ the relation

$$\text{(5-31)} \qquad x_p(iT) = N^{-1} \sum_{k=0}^{N-1} \tilde{X}\left(k\frac{2\pi}{NT}\right) e^{jik2\pi/N},$$

representing the inverse discrete Fourier transform.

5.3 Discrete Fourier transforms and interpolation

5.3.1 Interpolation in the image

For the Fourier transform of a symbolic expression of a sequence of function values x_i, $i = 0, 1, \ldots, N-1$, of a function x we have relation (5.24a). Replacing x_i in that from the relation for the inverse discrete Fourier transform, we get an interpolation relation

$$(5\text{-}32) \qquad \tilde{X}(\omega) = \sum_{i=0}^{N-1} x_i \, \mathrm{e}^{-\mathrm{j}\omega iT} = \frac{1}{N} \sum_{k=0}^{N-1} X_k \sum_{i=0}^{N-1} \mathrm{e}^{-\mathrm{j}i(\omega T + k2\pi/N)}$$

which leads to the relation

$$\tilde{X}(\omega) = \frac{1}{N} \sum_{k=0}^{N-1} X_k \frac{1 - \mathrm{e}^{-\mathrm{j}\omega TN}}{1 - \mathrm{e}^{-\mathrm{j}(\omega T - \varkappa 2\pi/N)}}, \qquad \omega \neq \varkappa \frac{2\pi}{NT}.$$

There is, however, a more suitable procedure for the interpolation linked with § 4.3.1 on padded sequences. Let us substitute $\omega = 2\pi/(N'T)$, $N' > N$, $\varkappa = 0, 1, \ldots, N' - 1$ from relation (5-24a); we get

$$(5\text{-}33) \qquad \tilde{X}\left(\varkappa \frac{2\pi}{N'T}\right) = \sum_{i=0}^{N-1} x_i \, \mathrm{e}^{-\mathrm{j}i\varkappa 2\pi/N'} .$$

Now, on introducing the padded sequence y_i to the sequence x_i

$$(5\text{-}34) \qquad y_i = \begin{cases} x_i, & i = 0, 1, \ldots, N-1, \\ 0, & i = N, \ldots, N'-1, \end{cases}$$

we can write (5-33) as follows:

$$\tilde{X}\left(\varkappa \frac{2\pi}{N'T}\right) = \sum_{i=0}^{N'-1} y_i \, \mathrm{e}^{-\mathrm{j}i\varkappa 2\pi/N'} = \mathscr{D}_{N'}\{y_i\} = Y_\varkappa .$$

Hence, one can compute $X(\varkappa . 2\pi/(NT)$, for all values $\omega = \varkappa . 2\pi/(N'T)$, $\varkappa = 0, 1, \ldots, N'-1$, $N' > N$, so that we compute the discrete Fourier transform of the sequence x_i padded with zero terms from N to N'.

By choosing $N' = 2N$ we can, after padding the sequence x_i with zero terms to double length, calculate $\tilde{X}(\omega)$ for $\omega = \varkappa\pi/(NT)$, $\varkappa = 0, 1, \ldots, 2N-1$, i.e. we can calculate $\tilde{X}(\omega)$ with a half step. At the same time $X_k = Y_{2k}$ for $k = 0, 1, \ldots, N-1$.

This trick is very often used in computing discrete Fourier transforms.

Example 17. If we carried out in example 14 a computation of the discrete Fourier transform e.g. for $a = 1$, $N = 128$ and $T = 0.0625$, i.e. $TN = 8$, then values TX_k would give approximate values $X_1(\omega)$ for $\omega = \omega_k = k\pi/4$, $k = 0, 1, \ldots, N/2 - 1$. If we need to obtain values $X_1(\omega)$ with a smaller step, such as 0.1, i.e. we need values $X_1(\omega)$ for $\omega \approx \varkappa . 0.1$, $\varkappa = 0,1, \ldots,$ then it is necessary that $2\pi/(N'T) \approx 0.1$. It follows that it is sufficient to choose $N' = 1024$. Thus we must pad a sequence x_i, $i = 0, 1, \dots, 127$, with 896 zero terms to get a sequence y_i, $i = 0, 1, \dots, 1024$. The discrete Fourier transform $Y_\varkappa$, $\varkappa = 0, 1, \dots, N' - 1$ of this sequence then approximately expresses values $X_1(\omega_\varkappa)$ for $\omega_\varkappa = \varkappa \,.\, 2\pi/(N'T) = \varkappa \,.\, 9.8 \,.\, 10^{-2}$, $\varkappa = 0, 1, \dots, N'/2 - 1$.

5.3.2 Interpolation in the original

We start with expression (5-11) for an interpolation polynomial

$$u(t) = \sum_{k=-M}^{M} D_k \, e^{jk2\pi t/P} .$$

We replace D_k from (5-14) either for the case $N = 2M$, i.e. N even, or for $N = 2M + 1$, i.e. N odd, to define the interpolation. Hence we obtain the interpolation relations

$$\text{(5-35a)} \quad u(t) = \frac{1}{N} \sum_{i=0}^{N-1} x_i \frac{\sin \pi \dfrac{t - iT}{T}}{\sin \pi \dfrac{t - iT}{NT}} , \quad N \text{ odd},$$

$$\text{(5-35b)} \quad u(t) = \frac{1}{N} \sum_{i=0}^{N-1} x_i \frac{\sin \pi \dfrac{t - iT}{T}}{\sin \pi \dfrac{t - iT}{NT}} \cos \pi \frac{t - iT}{NT} , \quad N \text{ even}.$$

One can, however, use the inverse Fourier transform for the interpolation, starting from (5-11) where we substitute from (5-14) and so getting for

Table 4.

N \ N'	even	odd
even	$N' = N + 2\nu$ $Y_k = \begin{cases} X_k, \ k = 0, 1, \dots, \dfrac{N}{2}, \\ 0, \ \ k = \dfrac{N}{2}, \dots, \dfrac{N}{2} + \nu \end{cases}$	$N' = N + 2\nu + 1$ $Y_k = \begin{cases} X_k, \ k = 0, 1, \dots, \dfrac{N}{2}, \\ 0, \ \ k = \dfrac{N}{2} + 1, \dots, \dfrac{N}{2} + 1 + \nu \end{cases}$
odd	$N' = N + 2\nu + 1$ $Y_k = \begin{cases} X_k, \ k = 0, 1, \dots, \dfrac{N - 1}{2} \\ 0, \ \ k = \dfrac{N + 1}{2}, \dots, \dfrac{N + 1}{2} + \nu \end{cases}$	$N' = N + 2\nu$ $Y_k = \begin{cases} X_k, \ k = 0, 1, \dots, \dfrac{N - 1}{2}, \\ 0, \ \ k = \dfrac{N + 1}{2}, \dots, \dfrac{N + 1}{2} + \nu \end{cases}$

a real interpolation polynomial the form

$$(5\text{-}36a)\quad u(t) = \frac{1}{N}[X_0 + 2\,\mathrm{Re} \sum_{k=1}^{(N-1)/2} X_k\, e^{jkt2\pi/(NT)}], \quad N \text{ odd},$$

$$(5\text{-}36b)\quad u(t) = \frac{1}{N}\left[X_0 + 2\,\mathrm{Re} \sum_{k=1}^{N/2-1} X_k\, e^{jkt2\pi/(NT)} + X_{N/2} \cos \frac{\pi}{T} t\right], \quad N \text{ even}.$$

Relations (5-36) turn into the inverse discrete Fourier transform for $t = iT$. If we want to compute $u(t)$ for $t = lNT/N'$, $N' > N$, $l = 0, 1, \ldots, N' - 1$, it is enough to compute the inverse discrete Fourier transform of a sequence Y_k, that arises from X_k by padding with zero terms in such a way that it is again a discrete Fourier transform of a real sequence. Creation of the sequence Y_k from the sequence X_k is described in Table 4.

5.4 Computation of a Fourier transform by means of the discrete Fourier transform

5.4.1 Use of the discrete Fourier transform to calculate a Fourier transform by Eagle's method

In Chapter 3 we described Eagle's method of calculating a Fourier transform. From expression (3-29) it follows for $\omega = k\,.\,2\pi/(NT)$ and $t = iT$ that:

$$(5\text{-}37)\quad X\left(k\frac{2\pi}{NT}\right) = \sum_{\nu=0}^{n}\left(jk\frac{2\pi}{NT}\right)^{-\nu-1}[\sum_{i=0}^{N-1} \Delta_1^{(\nu)}\, e^{-jik2\pi/N} + \Delta_N^{(\nu)}].$$

In the general case it would therefore be necessary to carry out a calculation of the discrete Fourier transform of the sequence $\Delta_i^{(\nu)}$ for $\nu = 0, 1, \ldots, n$, i.e. altogether $n + 1$ calculations of the discrete Fourier transform.

In the case where $n = 1$ or $n = 0$, we get the relations needed from the relations for the equidistant case from § 3.2.2. For the equidistant spline interpolation, general relations may also be used, but in the next section a simpler method will be presented.

5.4.2 Use of spline interpolation and the discrete Fourier transform for calculating a Fourier transform

Let us assume that the function x is expressed by an equidistant spline with the inner nodes $x(iT) = x_i$, $i = 1, 2, \ldots, N - 1$, and with the outer nodes $x(0) = x_0$, $x(NT) = x_N$. Functions $\varphi_i(t)$ are polynomials of n-th order subject to the spline conditions

$$\varphi_i^{(\nu)}(iT) = \varphi_{i-1}^{(\nu)}(iT), \qquad \nu = 0, 1, \ldots, n-1$$
$$i = 1, \ldots, N-1.$$

Then relation (3-42b) holds for the Fourier transform by (3-29), from which we get

$$\text{(5-38)} \quad X(\omega) = \sum_{\nu=0}^{n-1} (\mathrm{j}\omega)^{-\nu-1} [\Delta_0^{(\nu)} - \Delta_N^{(\nu)} \mathrm{e}^{-\mathrm{j}\omega iT}] + $$
$$+ (\mathrm{j}\omega)^{-n-1} \sum_{i=0}^{N} \Delta_i^{(n)} \mathrm{e}^{-\mathrm{j}\omega iT}.$$

Let us assume that

$$\text{(5-39)} \quad \Delta_0^{(\nu)} = \Delta_N^{(\nu)} = 0, \qquad \nu = 2, 3, \ldots, n-1,$$

and introduce an additional assumption

$$\text{(5-40)} \quad \Delta_0 = \Delta_N = 0, \qquad \Delta_0^{(1)} = \Delta_N^{(1)} = 0.$$

Since $\Delta_0^{(2)} = \Delta_N^{(2)} = 0$, this just means for a natural spline of the third order the introduction of the additional assumptions (5-40).

Using these conditions the expression for $X(\omega)$ simplifies to the form

$$\text{(5-41)} \quad X(\omega) = (\mathrm{j}\omega)^{-n-1} \sum_{i=0}^{N} \Delta_i^{(n)} \mathrm{e}^{-\mathrm{j}\omega iT}.$$

Let us now use expression (5.24b) for the Fourier transform of a symbolically expressed sequence of function values $x(iT) = x_i$

$$\text{(5-42)} \quad \tilde{X}(\omega) = \frac{1}{T} \sum_{\nu=-\infty}^{\infty} X\left(\omega - \nu \frac{2\pi}{T}\right).$$

Now substituting for $X(\omega)$ from (5-41) into (5-42) we get

$$\text{(5-43)} \quad \tilde{X}(\omega) = \frac{1}{T} \sum_{\nu=-\infty}^{\infty} \frac{1}{(\mathrm{j}(\omega - \nu \,.\, 2\pi/T))^{n+1}} \sum_{i=0}^{N} \Delta_i^{(n)} \mathrm{e}^{-\mathrm{j}\omega iT}.$$

If we express the finite sum in this relation by (5-41), we get the relation between $X(\omega)$ and $\tilde{X}(\omega)$:

$$\text{(5-44)} \quad X(\omega) = TA_n(\omega)\, \tilde{X}(\omega),$$

where

$$\text{(5-45)} \quad A_n(\omega) = \frac{1}{\omega^{n+1}} \frac{1}{\sum\limits_{\nu=-\infty}^{\infty} \dfrac{1}{(\omega - \nu \,.\, 2\pi/T)^{n+1}}}.$$

Thus we arrive at the important conclusion that the Fourier transform of the function with the given spline properties can be obtained from the Fourier transform of the symbolic expression of their values $x(iT)$ by means of the coefficient $A_n(\omega)$. The expression for $A(\omega)$ can further be modified using relations

$$\sum_{\nu=-\infty}^{\infty} \frac{1}{x - \nu u} = \frac{\pi}{u} \operatorname{cotg} \pi \frac{x}{u},$$

$$\frac{1}{x^n} = \frac{(-1)^{n-1}}{(n-1)!} \left(\frac{1}{x}\right)^{(n-1)},$$

which lead to the relation

$$\text{(5-46)} \qquad \sum_{\nu=-\infty}^{\infty} \frac{1}{(x - \nu u)^n} = \frac{\pi}{u} \frac{(-1)^{n-1}}{(n-1)!} \left(\operatorname{cotg} \pi \frac{x}{u}\right)^{(n-1)},$$

to a form

$$\text{(5-47)} \qquad A_n(\omega) = \frac{\dfrac{\mathrm{d}^n}{\mathrm{d}\omega^n}\left(\dfrac{2}{\omega T}\right)}{\dfrac{\mathrm{d}^n}{\mathrm{d}\omega^n} \operatorname{cotg} \dfrac{\omega T}{2}}.$$

Then for $n = 1$, 3 and 5 we get

$$A_1(\omega) = \left(\frac{\sin \omega T/2}{\omega T/2}\right)^2,$$

$$A_3(\omega) = \frac{3}{2 + \cos \omega T} \left(\frac{\sin \omega T/2}{\omega T/2}\right)^4,$$

$$A_5(\omega) = 60 \left(\frac{\sin \omega T/2}{\omega T/2}\right)^6 (11 \sin^2 \omega T/2 + $$
$$+ 3 \sin 3\omega T/2 \,.\, \sin \omega T/2 + 55 \cos^2 \omega T/2 +$$
$$+ 5 \cos 3\omega T/2 \,.\, \cos \omega T/2)^{-1}.$$

A sketch of the weight function $A_n(\omega)$ is presented in Figure 17.

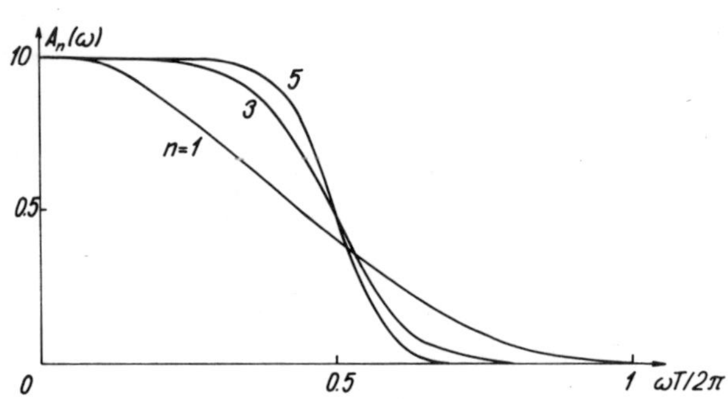

17. Curve of the weight function A_n

If we compare this procedure with the procedure of Chapter 3 for an equidistant spline, it is then obvious that the current procedure is much simpler. Introducing the additional conditions of zero values of the function and its derivative for $t = 0$ and $t = NT$ brings a substantial simplification in the case of the third order spline. The given conditions can be satisfied by subtracting a suitable function from the function x as was shown in § 3.3.2.

The expression of relations between $X(\omega)$ and $\tilde{X}(\omega)$ is not new. It was first described in 1959 by F. L. Bauer and H.J. Stetter [12] and interpreted in a somewhat different way by W. Gauschi [41], [83] and D. Achilles [1].

6. METHODS OF COMPUTATION OF DISCRETE FOURIER TRANSFORMS

6.1 Introduction

The computation of the discrete Fourier transform of a sequence of N values x_i represents a computation of N terms of the sequence X_k. It follows from (4-1) that it is necessary to carry out N^2 complex multiplications and $N(N - 1)$ complex additions. The quadratic dependence of the number of computational operations on N already prompted the search for more efficient computing procedures at applications of Bessel relations in harmonic analysis. The Runge method [74], [76], first mentioned in 1905, and using forms, was the first and for a long time only efficient method. The next step was not until the publication of a new algorithm for computing discrete Fourier transforms in 1965 by J. W. Cooley and J. W. Tukey: the fast Fourier transform (FFT). It appeared later (see e.g. [21]) that a number of authors had earlier come very close to the fast Fourier transform but their methods had not become widespread.

Algorithms for the fast computation of discrete Fourier transforms became a matter of intensive interest after 1965 and were presented in a wide variety of modifications. For everyday purposes, they may be considered equivalent to computer computations. The difference, however, is significant when designing the inner configuration of dedicated devices for their computation, i.e. FFT processors.

The effort to speed up the computation has led to a search for other algorithms. As a result, Winograd algorithm was set up in 1976 [78]. In addition, other algorithms for computing discrete Fourier transforms were found, which were not so fast, but were suitable for special applications. Examples of these are Goertzel and Bluestein algorithms. These will also be described, but first we describe the principle of fast algorithms for the case $N = 2^m$ and then for the general case.

The principle of FFT will be explained in an elementary manner and illustrated by means of graphs. Interpretation can, of course, be carried out purely algebraically by means of sums [25], [17] or matrices [69], [82] and their Kronecker products [46].

6.2 Algorithms for the fast Fourier transform for $N = 2^m$

6.2.1 An algorithm for the fast Fourier transform by the method of selected input sequences

This algorithm was created by J. W. Cooley and J. W. Tukey and is usually called *a decimation-in-time algorithm* (DIT) [20].

Let us split a sequence $x_i \triangleq X_k$ of N terms into two selected subsequences of $N/2$ terms, one of even terms, the other of odd terms. This procedure is usually called a fast Fourier transform with base 2. The transform X_k can then be written as follows:

$$(6\text{-}1)\qquad X_k = \sum_{i=0}^{N-1} x_i\, e^{-jik2\pi/N} = \sum_{v=0}^{N/2-1} x_{2v}\, e^{-jvk2\pi/(N/2)} + \\ + e^{-jk2\pi/N} \sum_{v=0}^{N/2-1} x_{2v/1}\, e^{-jvk2\pi/(N/2)}, \qquad k = 0, 1, \ldots, N-1.$$

Denoting the transform of the sequence of even terms

$$(6\text{-}2)\qquad \mathscr{D}_{N/2}\{x_{2v}\} = G_k, \qquad k = 0, 1, \ldots, N/2 - 1,$$

and the transform of the sequence of odd terms

$$(6\text{-}3)\qquad \mathscr{D}_{N/2}\{x_{2v+1}\} = H_k, \qquad k = 0, 1, \ldots, N/2 - 1,$$

the transform of the original sequence can be expressed in the form

$$(6\text{-}4)\qquad X_k = G_k + e^{-jk2\pi/N} H_k, \qquad k = 0, 1, \ldots, N-1,$$

which holds for given k with respect to periodicities H_k and G_k. This relation may be written as two relations:

$$(6\text{-}5)\qquad X_k = G_k + W^k H_k, \qquad k = 0, 1, \ldots, N/2 - 1,$$

$$(6\text{-}6)\qquad X_{k+N/2} = G_k - W^k H_k, \qquad k = 0, 1, \ldots, N/2 - 1,$$

where $W = e^{-j2\pi/N}$, and $W^{k+N/2} = -W^k$.

The computation of X_k from G_k and H_k can be shown graphically by means of a signal graph. Rules for setting up the signal graph are shown in the upper part of Figure 18. In the bottom part, the corresponding signal graph is drawn on the basis of these rules, together with an equivalent graph which contains only one multiplication by the complex quantity $W^k = e^{-jk2\pi/N}$. This elementary graph is usually called a "butterfly".

Relations (6-5) and (6-6) can also be obtained from relations (4-118) and (4-119). In addition, one can show that (6-5) and (6-6) express the discrete Fourier transform for $N = 2$ from values G_k and $W^k H_k$. The whole computation procedure is shown in Figure 19 for $N = 8$. Further computing is required to carry out the discrete Fourier transform of

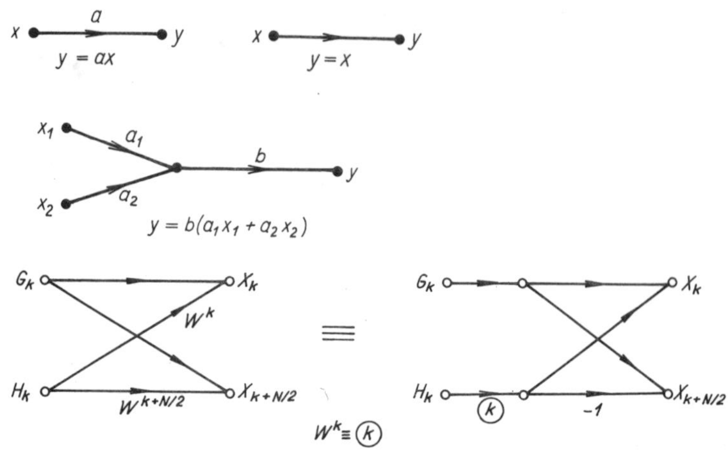

18. Graphical illustration of the computation X_k from G_k and H_k by means of a signal graph

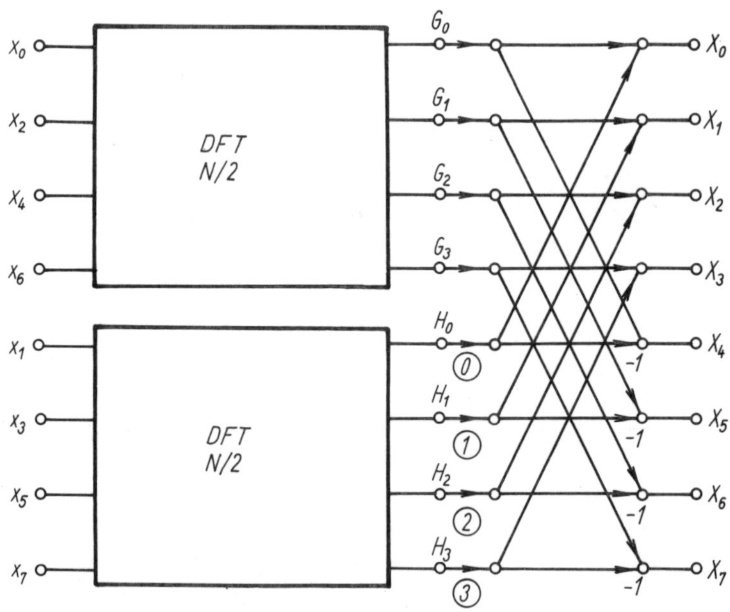

19. First step in the computation of the discrete Fourier transform for $N = 8$ by decimation in time

a sequence of even terms and of one of odd terms. This can be done in quite a similar manner, if we substitute $N/2$ instead of N. The result is shown in Figure 20. The last step is to carry out four transforms for $N = 2$, expressed directly by the "butterfly". The whole procedure is summed up in Figure 21. It is obvious from the figure that the image sequence is

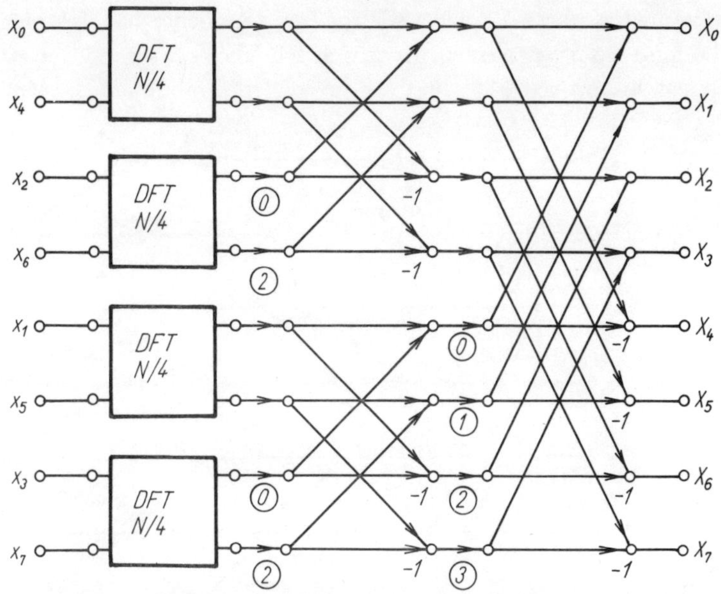

20. Second step in the computation of the discrete Fourier transform for $N = 8$ by decimation in time

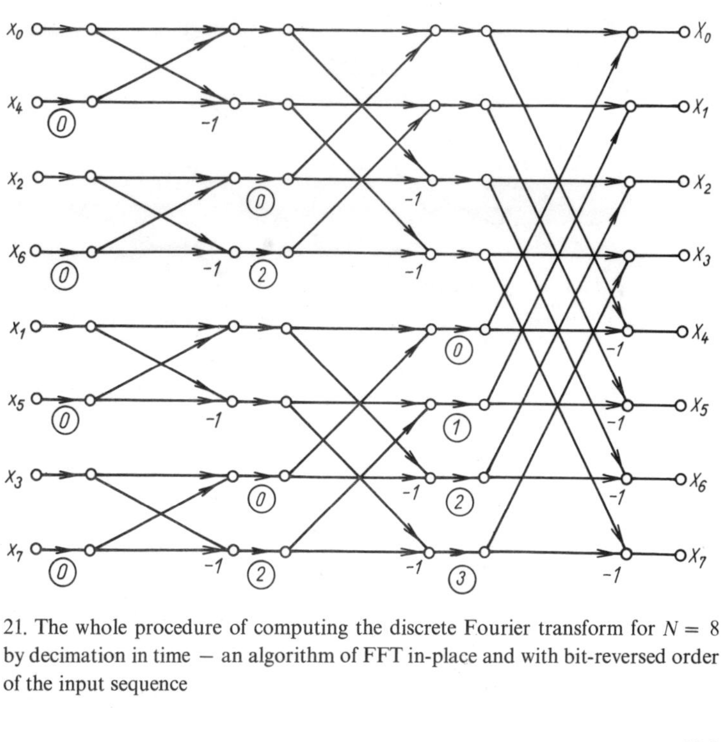

21. The whole procedure of computing the discrete Fourier transform for $N = 8$ by decimation in time – an algorithm of FFT in-place and with bit-reversed order of the input sequence

obtained in normal order while the input sequence has its order changed because we have selected from each sequence always odd or even terms. It can be proved [66] that for $N = 2^m$ the input sequence must be in

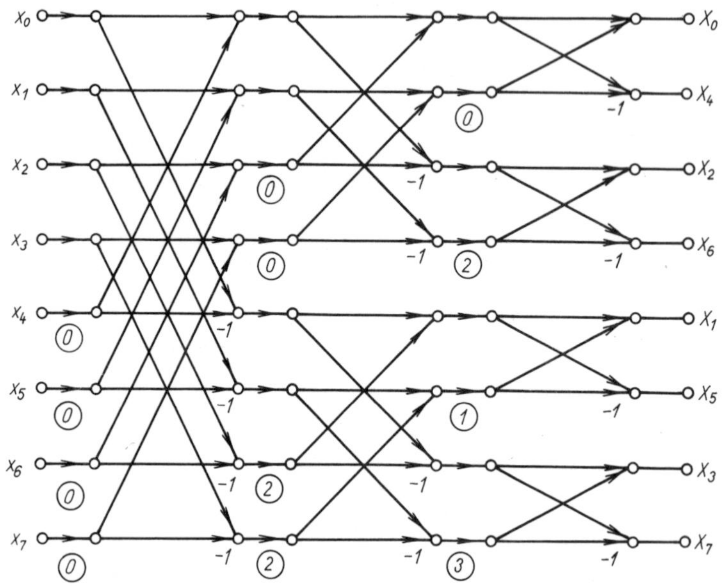

22. An algorithm of FFT in-place with a bit-reversed output sequence

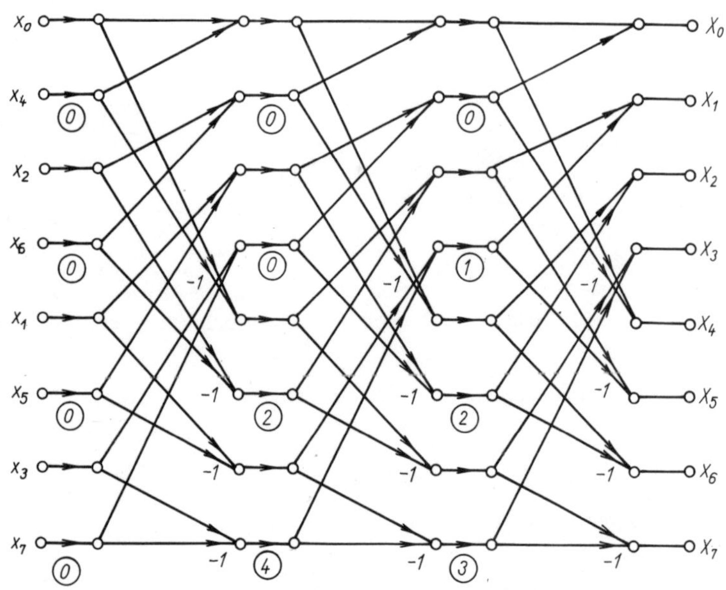

23. An algorithm of FFT with fixed geometry and a bit-reversed input sequence

a *bit-reversed order of indices.* The bit-reversed order is obtained from the normal order of indices by expressing the indices in binary form and reading each binary number backwards [66]. So for example a number $k = 3$ with binary expression 011 becomes 110 with corresponding decimal number 6.

A graphical representation for $m > 3$ can be derived by further repeating this procedure. The branching method of a graph does not change, but the weight coefficients W^k of branches do vary. It is a characteristic of this method of computation that the results of computations contained in a certain row do not need to be removed into another row for a computation in the next column. This algorithm is therefore called *in place.*

The algorithm of Figure 21 can be imagined with the values X_k in the bit-reversed order. We would then get the graph shown in Figure 22 that again corresponds to a procedure without removing. Another way of charting Figure 21 allows us to set up the graph shown in Figure 23. In the case of this algorithm, it is characteristic that all columns possess the same geometry and it is therefore called *an algorithm with a fixed geometry with a bit-reversed input sequence.* Another method of computing an FFT can be devised, however, in which neither the input nor the image sequence is bit-reversed. The algorithm with these properties is shown in Figure 24. In total, the method of selected input sequences has produced four different graphs for this algorithm.

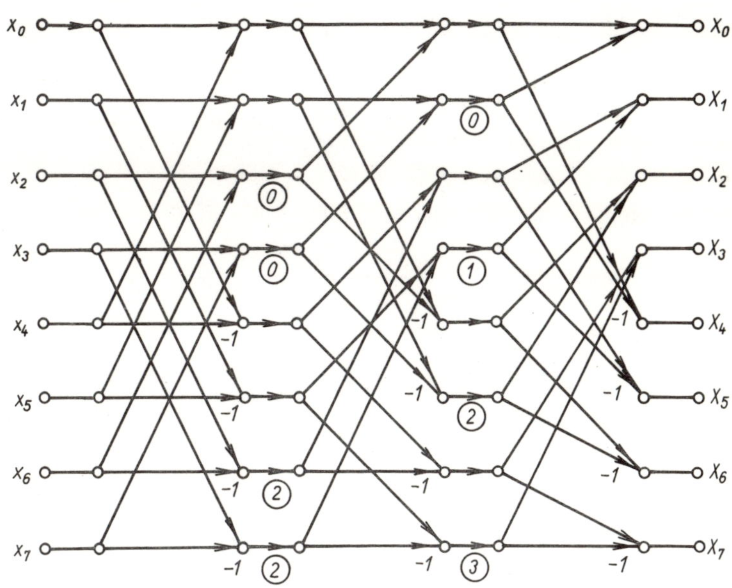

24. An algorithm of FFT with variable geometry and normal order in an input, as well as in an output

6.2.2 An algorithm for the fast Fourier transform by the method of selected output sequences

This method is sometimes called also *a decimation-in-frequency* (DIF) *algorithm* [43]. The principle of computation consists in independently computing the even and the odd terms of the output sequence. For even terms of the transform with $k = 2\mu$, $\mu = 0, 1, \dots, N/2 - 1$, the defining relation leads to the expression

$$(6\text{-}7) \qquad X_{2\mu} = \sum_{i=0}^{N-1} x_i \, e^{-ji2\mu 2\pi/N} = \sum_{i=0}^{N-1} x_i \, e^{-ji\mu 2\pi/(N/2)} =$$
$$= \sum_{i=0}^{N/2-1} x_i \, e^{-ji\mu 2\pi/(N/2)} + \sum_{i=0}^{N/2-1} x_{i+N/2} \, e^{-ji\mu 2\pi/(N/2)} =$$
$$= \mathscr{D}_{N/2}\{x_i + x_{i+N/2}\}.$$

Similarly, for the odd terms of the transform we get

$$(6\text{-}8) \qquad X_{2\mu+1} = \mathscr{D}_{N/2}\{(x_i - x_{i-N/2})\, e^{-ji2\pi/N}\}.$$

After introducing ancillary sequences

$$x_i + x_{i+N/2} = r_i, \qquad (x_i - x_{i+N/2})\, W^i = s_i, \qquad i = 0, 1, \dots, N/2 - 1,$$

we can express transforms of these sequences by the relations

$$(6\text{-}9) \qquad X_{2\mu} = \mathscr{D}_{N/2}\{r_i\}, \quad X_{2\mu+1} = \mathscr{D}_{N/2}\{s_i\}.$$

The computation of the quantities r_i and s_i can again be represented graphically in the form of a "butterfly" (Figure 25). It follows from relations (6-9) that they convert computation of the transform of a sequence of N values to a computation of the transform of sequences r_i and s_i of half the length. The computation for all X_k is shown in Figure 26 for $N = 8$.

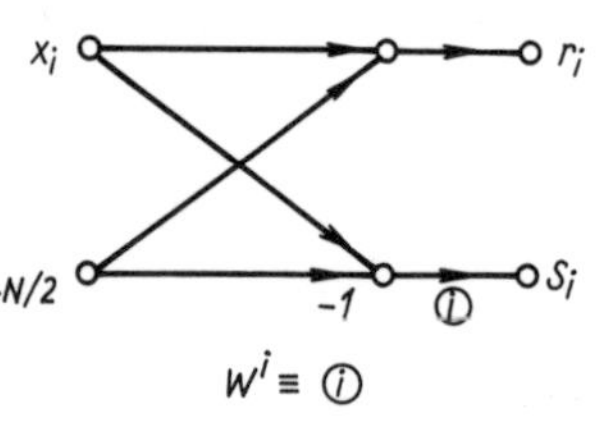

25. Graphical illustration of the computation of r_i and s_i by means of a signal graph

Repeating the same procedure for the computation of a transform from $N/2 = 4$ values leads to Figure 27. Further repetition of the procedure for $N/4 = 2$ leads to the graph of Figure 28. In this case the image sequence is bit-reversed and the algorithm again runs in place. This algorithm can further be modified in various ways. Algorithms based on selected input

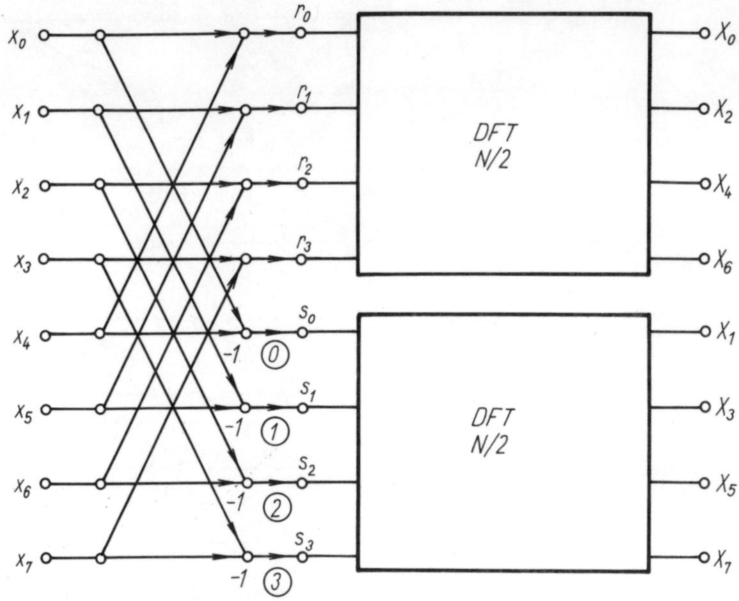

26. First step in the computation of the discrete Fourier transform for $N = 8$ by decimation in frequency

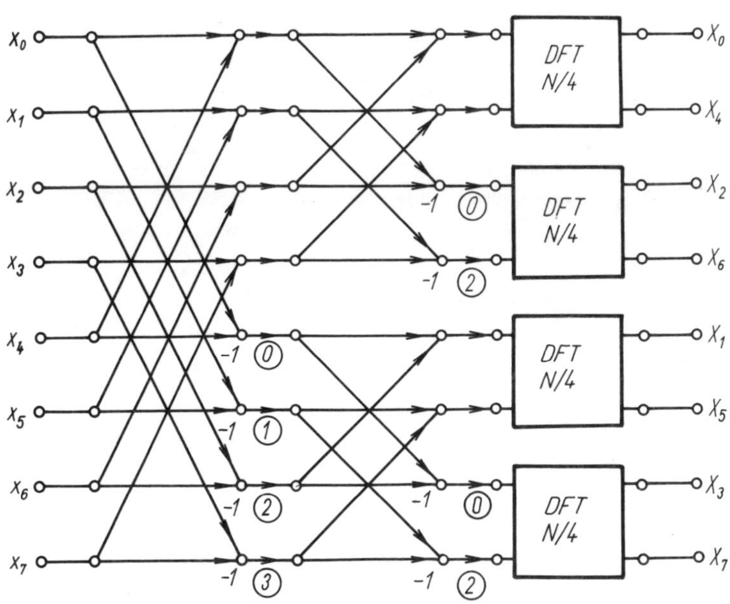

27. Second step in the computation of the discrete Fourier transform for $N = 8$ by decimation in frequency

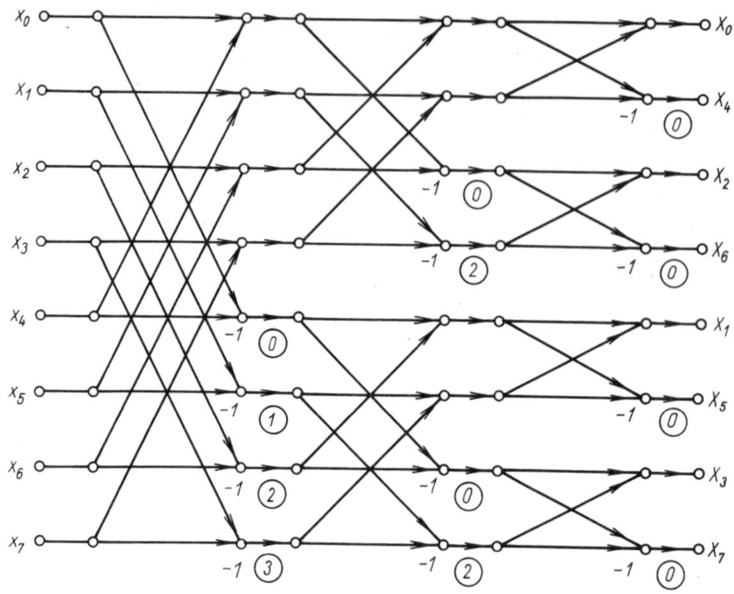

28. The whole procedure of computing the discrete Fourier transform by decimation in frequency — an algorithm of FFT in place with bit-reversed order of an output sequence

sequences and algorithms based upon selected output sequences are mutually inverse [66]. An inversion in this case means that the graph of an algorithm based upon the selected output sequence is obtained from the graph of an algorithm based upon the selected input sequences (and vice versa) by changing the directions of all arrows while leaving the coefficients W and reading the graph from right to left. We can easily find out that the graph in Figure 28 is inverse to the graph in Figure 21. This way three other algorithms with selected output sequences may be constructed by inversion of the graphs of Figures 22, 23 and 24.

It is therefore possible to set up altogether eight basic algorithms for the FFT.

Hitherto all algorithms were derived just for the direct discrete Fourier transform. For the inverse transform, it is not necessary to establish an independent algorithm because it is possible to apply theorem 2 of § 4.1.2 or to replace x_i in the algorithms by the sequence $N^{-1}X_k$ and to put W^* instead of W.

6.2.3 Speed of computation and choice of algorithm

As has already been shown in § 6.1, it is necessary in the case of computation by the defining relation to carry out N^2 complex multiplications and

$N(N-1)$ complex additions. The "butterflies" and algorithms described prove that in the case of FFT one complex multiplication and two complex additions are necessary in each "butterfly", while we do not consider multiplication by -1 as a multiplication. From the geometry of the graphs it follows that for $N = 2^m$ each algorithm is set from $mN/2$ "butterflies" and hence in total it contains $mN/2$ multiplications and Nm additions.†)

When we compare the number of multiplications in a computation by the defining relation with an algorithm of the Fourier transform, we get for the coefficient α_m, giving the reduction of the number of multiplications:

$$\alpha_m = \frac{N^2}{(N/2)\,m} = \frac{2N}{m}.$$

It is obvious from the relation that the number of multiplications is reduced more than a hundred times for sequences of length $N = 512$ and more than a thousand times for sequences of length $N = 8192$. A substantial reduction in the number of multiplications, especially for large N, has made it possible to carry out approximate computations of the Fourier transform by means of the FFT algorithms to an extent that would be unacceptably time-consuming, even on the fastest computers, using the classical computing methods.

Apart from the speed, it is important to know also the precision of computing the discrete Fourier transform. The precision of computation depends first of all on the wordlength used by the computer. In the algorithms of FFT a common practice has been to express the precision of computation by the ratio of the variance σ_E^2 of the error sequence $|\hat{X}_k - X_k|$ to the variance σ_x^2 of the output sequence X_k. Here the quantity X_k means a precise result, i.e. the result at a theoretically infinite wordlength, and $\hat{X}_k$ the result at a finite wordlength. Denoting the ratio of the variances by $\beta = \sigma_E^2/\sigma_x^2$, one can obtain for it the relation [66]

$$\beta \approx 0.8m\, 2^{-2b}$$

valid for an algorithm with $N = 2^m$ when the wordlength of the mantissa is b binary digits and a random rounding in the floating point is used. For the case of non-random rounding, then [66]

$$\beta \approx 0.32m^2\, 2^{-2b}\,.$$

†) If we consider the number of multiplications before the simplification of "butterflies", then two multiplications and two additions are necessary in a "butterfly", i.e. altogether four operations. For $N/2$ this represents $2N$ operations. For the number of operations in the general case see § 6.3.2.

There arises the question of which of the described algorithms is most suitable. They all contain the same number of computing operations and from this point of view are all equally suitable. The computation time depends, however, not only on the number of computing operations but also on the speed at which the computer performs multiplication and addition. From the point of view of speed the most efficient programs are those which are maximally adapted to the virtues of the individual computer, thus usually programs which are part of the software provided by the computer manufacturer. If no such programs are available, it will generally be satisfactory to use the versatile programs presented e.g. in [17], [22], [66], [71], [4], [15]. Only in a case where there are special requirements on precision or speed will there be any reason to make new programs.

6.3 Fast Fourier transform for general N

6.3.1 Fast Fourier transform of the base $N = r^n$

For the algorithms described in the last paragraph the name base-2 fast Fourier transform is usually used. This whole procedure can be generalized to any base and so it is possible to set up algorithms for $N = r^m$. Of practical significance are algorithms of base $r = 4$ and $r = 8$.

The principle of such an algorithm is again a construction of elementary geometry (an analogy of the "butterfly"). This geometry follows from the procedure when we successively select from a sequence of N terms each fourth term beginning with the term with zero index. So we make four selected subsequences of length $N/4$ with general term

$$x_{4\nu+M}, \qquad \nu = 0, 1, \ldots, N/4 - 1, \quad m = 0, 1, 2, 3.$$

From relation (4-121), which expresses the transform of a whole sequence by means of the transform of selected sequences for $p = 4$ and $M = N/4$, we get

$$X_{\mu N/4+n} = \mathscr{D}_4\{Z_{m,n}\} = \sum_{m=0}^{3} Z_{m,n}\, \mathrm{e}^{-jm\mu 2\pi/4} \qquad \mu = 0, 1, 2, 3, \quad n = 0, 1, \ldots, N/4,$$

where

$$Z_{m,n} = X_{m,n}\, \mathrm{e}^{-jmn2\pi/N}$$

and

$$X_{m,n} = \mathscr{D}_{N/4}\{x_{4\nu+m}\}.$$

The advantage of this procedure is that it uses a computation of the discrete Fourier transform for $N = 4$ where only the coefficients 1, -1, j and $-$j occur, and therefore does not need to do multiplications. The only multiplication that must be done is given by the exponential coefficient in the expression for $Z_{m,n}$.

For the computation of a discrete Fourier transform of bases 4 or 8 very effective programs can be set up which will be discussed in § 6.3.2. Base-4 algorithms are also used widely in dedicated processors for fast computation of the discrete Fourier transform.

For the base-4 algorithm a relation can be derived for the number of complex multiplications without a multiplication by one arising from the expression W^0. If $N = 2^m = 4^{m/2}$, then the number of multiplications at $r = 4$ is

$$\gamma_4 = \tfrac{3}{4}mN - (N - 1)$$

and at $r = 2$

$$\gamma_2 = \tfrac{1}{2}mN - (N - 1).$$

Thus, for instance, for $N = 1024$ and $r = 2$, $\gamma_2 = 4097$ multiplications are necessary, while at $r = 4$ only $\gamma_4 = 2817$ multiplications are needed.

The base-8 algorithm presents a further possibility of reducing the number of multiplications, but hardly by very much.

6.3.2 Fast Fourier transform with a composite base

Cooley and Tukey [25] also formulated a fast computation algorithm for $N = r_i r_2 \ \dots \ r_m$. We first demonstrate the essence of this procedure for $N = 3.2$. In this case the sequence x_i can again be split into a sequence of even terms x_0, x_2, x_4 and a sequence of odd terms x_1, x_3, x_5. Applying § 4.3.4, we create from a sequence of $N = 6$ terms $p = 2$ sequences of $M = 3$ terms so that we take every p-th one beginning with the zero-th and every p-th beginning with the first. From relation (4-121) we then get

$$X_{3\mu+n} = \sum_{m=0}^{1} e^{-jmn2\pi/6} \left(\sum_{v=0}^{2} x_{2v+m}\, e^{-jvn2\pi/3}\right) e^{-jm\mu 2\pi/2}. \tag{6-10}$$

First we carry out the inner addition, i.e. the transform for $N = 3$ and $m = 0$, representing the transform of the even terms and afterwards for $m = 1$, representing the transform of the odd terms. We get intermediate results represented by two sequences ${}_0X_n$ and ${}_1X_n$, $n = 0, 1, 2$:

$${}_mX_n = \sum_{v=0}^{2} x_{2v+m}\, e^{-jvn2\pi/3}, \qquad n = 0, 1, 2,$$
$$m = 0, 1.$$

We then multiply these results by terms $e^{-jmn2\pi/6}$ and we carry out the computation of the transform for $N = 2$ expressed by the outer addition. A graph obtained in this way is shown in Figure 29.

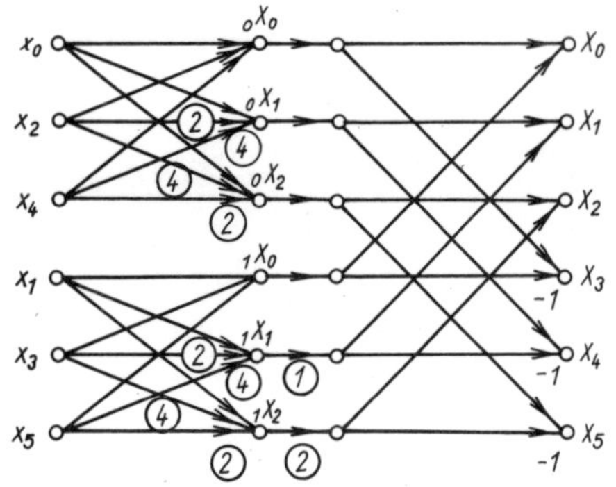

29. The procedure for the computation for $N = 6$, $p = 2$, $M = 3$

With the contrary choice $M = 2$, $p = 3$, we get from (4-121)

$$X_{2\mu+n} = \sum_{m=0}^{2} e^{-jmn2\pi/6} \left(\sum_{v=0}^{1} x_{3v+m} e^{-jvn2\pi/2} \right) e^{-jm\mu 2\pi/3} . \tag{6-11}$$

After denoting a discrete Fourier transform for $N = 2$

$$_mX'_n = \sum_{v=0}^{1} x_{3v+m} e^{-jvn2\pi/2}, \qquad n = 0, 1, \qquad m = 0, 1, 2, \tag{6-12}$$

we can write the result in the form of a discrete Fourier transform for $N = 3$

$$X_{2\mu+n} = \sum_{m=0}^{2} (e^{-jmn2\pi/6} {_mX'_n}) e^{-jm\mu 2\pi/3} . \tag{6-13}$$

This procedure is shown graphically in Figure 30.

The links between computations, i.e. the transform given by the inner addition and the one expressed by the outer addition, and the meaning of

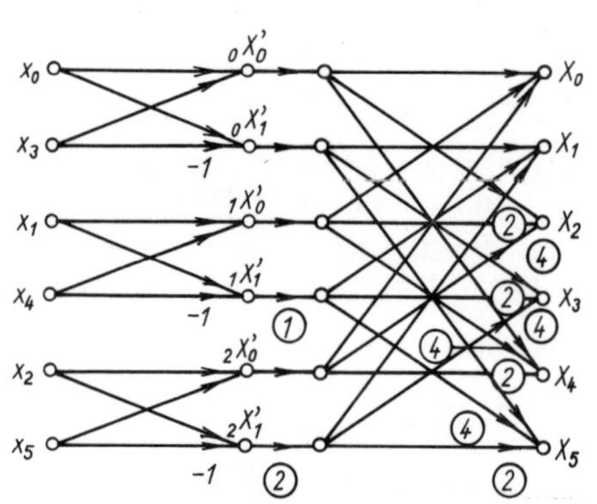

30. The procedure for the computation for $N = 6$, $p = 3$, $M = 2$

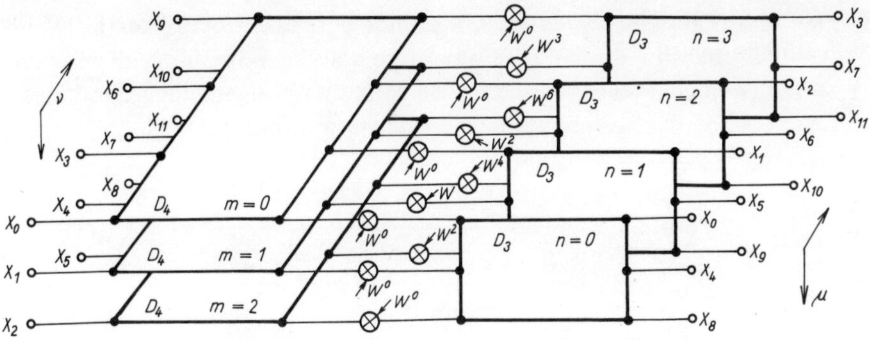

31. A spatial illustration of a computation of the discrete Fourier transform for $N = 12$, $p = 3$, $M = 4$

coefficients W_N^{mn} may be represented spatially [45]. This is done in Figure 31 for $N = 12$, taking $p = 3$ and $M = 4$. The figure may be interpreted as a computation by the algorithm of selected input sequences $p = 3$ taking the coefficients W^{mn} to the transform $\mathscr{D}_3$. We can also consider the spatial representation as an algorithm of selected output sequences $p' = 4$ when taking the coefficients W^{mn} to the transform $\mathscr{D}_4$. It is also obvious from the figure that it can be taken as a representation of the case $p = 4$, $M = 3$, if we read it from right to left.

From the above considerations it follows that in the general case, where

$$N = \prod_{\mu=1}^{m} r_\mu,$$

there will be needed

$$n_{op} = N \sum_{\mu=1}^{m} r_\mu$$

operations altogether, i.e. complex multiplications and additions. One can prove for the case $N = r^m$ that n_{op} has a minimum for $r = \mathrm{e}$. From this point of view it would be most suitable to perform FFT with the base $r = 3$. This applies to the case where we count as multiplications those involving $W_N^k = 1$ or $W_N^k = -1$. If we do not take these multiplications into account, then the most advantageous algorithms are for $r = 4$ and $r = 8$.

At the end of this paragraph we shall present some data about the CPU–time used by the IBM 370/135 computer using subroutines FFA and FFS that appear to be the fastest for sequences of length $N = 2^m$. The subroutine FFA computes the discrete Fourier transform of real sequences and the subroutine FFS the inverse transform again for the case of a real sequence. Both subroutines were taken over from [12] and were adjusted for double precision computation. Both subroutines work with a combined base 8, 4 and 2. The computing times achieved, obtained by averaging five computations, are shown in Table 5.

Table 5.

N	Average time in milliseconds			
	single precision		double precision	
	FFA DFT	FFS IDFT	FFA DFT	FFS IDFT
16	10	12	14	18
32	20	26	24	36
64	48	62	62	80
128	98	130	130	162
256	192	240	272	340
512	406	514	578	712
1 024	928	1 206	1 338	1 632
2 048	2 110	2 110	2 704	3 220
4 096	4 662	5 536	6 036	6 876

For work with complex sequences the subroutine FFT 842 presented in [12] is suitable. It computes about twice as slowly as FFA.

Singleton's program [80] and its newer version [81] are available for computing the discrete Fourier transform for an N composite. This program computes for composite numbers close to numbers 2^m about 3–5 times more slowly. Thus for instance a computation for $N = 126$ takes 368 (472) ms, for $N = 1008$ it takes 3028 (3754) ms and for $N = 5040$ a computation takes 21600 (24528) ms. Data in brackets apply to double precision for which the subroutine was also adapted.

6.4 Other principles of the fast Fourier transform computation

In this section we shall briefly describe some procedures for computing discrete Fourier transforms which differ from procedures based upon the Cooley–Tukey algorithm and the like. A common property of the algorithms further described is that they use the close relation between the discrete Fourier transform computation and the computation of convolution.

6.4.1 Fast Fourier transform for N prime

The algorithms described for FFT are inapplicable in the case where N is a prime number. C. M. Rader [73] demonstrated in 1968 that in this case the discrete Fourier transform computation could be converted into the computation of a convolution of a sequence of $N - 1$ terms. Since N is a prime, $N - 1$ must be a composite number. It follows from theorem 17 that a periodic convolution can be computed by means of the discrete Fourier transform. For the computation of a convolution we need to carry out in this case (see Chapter 7) the computation of a transform of length $N - 1$ that can always at least be theoretically done by the procedures presented in a previous section. C. M. Rader [73] proved his assertion by the technique of permuting the sequence x_i by a transform of the index i by a primitive root of the number N and using some other knowledge from the theory of numbers. The conversion of the computation of a discrete Fourier transform to the computation of a convolution may seem to be an unnecessary complication. It possesses, however, great theoretical significance and has brought up a new factor in the technique of efficient computation of the discrete Fourier transform.

6.4.2 The Winograd algorithm for the computation of the discrete Fourier transform

The conversion of the computation of the discrete Fourier transform to the computation of a convolution enables us to create algorithms for computing discrete Fourier transforms for small N with the minimum or near-minimum of multiplications. This was achieved on the basis of new methods of computing a convolution by means of the Cook–Toom algorithm [2], [61] and of so-called rectangular transforms.

The second fact that enables us to create efficient algorithms has been the discovery of a connection between the one-dimensional and the two-dimensional or multidimensional Fourier transforms that was pointed out by J. J. Good in 1958 [46]. Comparing relation (7-5) for the two-dimensional discrete Fourier transform with relation (4-121) expressing the computation of a transform by splitting up into selected sequences, we see that the relations differ just by the presence of an exponential factor in (4-121). Since one can arrive at (4-121) by splitting an original sequence of length N into a set of P sequences of length N/P (by selecting every P-th term beginning with the term with index zero), there arises the question of whether a more general transform of indices exists so that the exponential (twiddle) factor would drop out. Such transforms were successfully found and thus it was made possible to convert the discrete

Fourier transform computation for composite N, i.e. $N = N_1 N_2 \dots N_S$, into the successive computation of one-dimensional transforms of lengths $N_1, N_2, \dots, N_S$. This is, however, possible only when the factors $N_1, N_2, \dots, N_S$ are mutually prime. It is obvious that this method of computing the discrete Fourier transforms reduces the number of multiplications because all multiplications by the twiddle factors disappear.

Both these facts allow the setting up of a new algorithm known as the Winograd algorithm (WFTA – Winograd Fourier Transform Algorithm). The explanation of the essence of this algorithm is complex and requires some knowledge of number theory. A detailed discussion may be found in [61] and [65]. A description of the algorithm and Winograd algorithms for computing a transform for small N may be found in [78]. Reference [51] is devoted to the general conversion of the p-dimensional discrete Fourier transform into the q-dimensional one. Reference [64] is concerned with an analysis of the speed of computation by this algorithm. For comparison we present here some times required for computation on the IBM 370/135 computer by a program presented in [61] that was adapted also for double precision computing. Thus e.g. for $N = 126$, 450 (480) ms are needed, for $N = 1008$, 3940 (3930) ms and for $N = 5040$, 20200 (20800) ms are taken. (The times in brackets correspond to double precision.) The times needed are thus very close to the times shown for Singleton's program for composite N.

6.4.3 Recurrent methods for computing discrete Fourier transforms

A. Goertzel algorithm

The defining relation for the discrete Fourier transform may be interpreted as the computation of the function value of a polynomial at a given point. We set up, therefore, a polynomial, whose coefficients are the terms of a sequence x_i, $i = 0, 1, \dots, N-1$

$$X(s) = \sum_{i=0}^{N-1} x_i s^i . \tag{6-14}$$

Then to compute the value of the polynomial $X(s)$ at the point $s = e^{-j\varphi}$ one can use Horner recurrent algorithm expressed by the relations

$$y_i = x_{N-1-i} + e^{-j\varphi} \cdot y_{i-1}, \qquad i = 0, 1, \dots, N-1, \tag{6-15}$$
$$y_{-1} = 0.$$

Then there holds

$$X(e^{-j\varphi}) = y_{N-1} .$$

After adapting the polynomial $X(s)$ to the form

$$X(s) = s^{N-1} \sum_{i=0}^{N-1} x_i s^{i-(N-1)}, \tag{6-16}$$

we can compute $X(e^{-j\varphi})$ in the following manner

$$\begin{aligned} z_i &= x_i + e^{j\varphi} z_{i-1}, \\ z_{-1} &= 0, \\ X(e^{-j\varphi}) &= e^{-j(N-1)\varphi} z_{N-1}. \end{aligned} \tag{6-17}$$

The sequence feeds into relation (6-17) from the beginning, while it inputs into (6-15) from the end. The relations given are also valid for complex x_i. In the case where x_i is a real sequence the relations can be adapted so that in the recurrent relation we replace i by $i - 1$ and get e.g. from relation (6-15)

$$\begin{aligned} y_{i-1} &= x_{N-i} + e^{-j\varphi} y_{i-2}, \qquad i = 0, 1, \ldots, N-1, \\ y_{-2} &= 0. \end{aligned}$$

Multiplying this relation by $e^{j\varphi}$ and subtracting it from (6-15) leads to a recurrent relation of 2nd order

$$\begin{aligned} y_i - 2\cos\varphi y_{i-1} + y_{i-2} &= x_{N-i-1} - e^{-j\varphi} x_{N-i}, \\ y_{-1} = y_{-2} &= 0, \end{aligned} \tag{6-18}$$

that can be split into two relations

$$\begin{aligned} u_i - 2\cos\varphi u_{i-1} + u_{i-2} &= x_{N-i-1}, \\ u_{-1} = u_{-2} &= 0, \\ y_i = u_i - e^{j\varphi} u_{i-1}, &\qquad i = 0, 1, \ldots, N-1. \end{aligned} \tag{6-19}$$

For $X(e^{-j\varphi})$ then follows

$$X(e^{-j\varphi}) = y_{N-1} = u_{N-1} - e^{j\varphi} u_{N-2}. \tag{6-20}$$

By analogy we get from relation (6-17)

$$\begin{aligned} v_i - 2\cos\varphi v_{i-1} + v_{i-2} &= x_i, \qquad i = 0, 1, \ldots, N-1, \\ v_{-1} = v_{-2} &= 0, \\ z_i &= v_i - e^{-j\varphi} v_{i-1}, \end{aligned} \tag{6-21}$$

$$X(e^{-j\varphi}) = e^{-j(N-1)\varphi} z_{N-1} = e^{-j(N-1)\varphi} v_{N-1} - e^{jN\varphi} v_{N-2}. \tag{6-22}$$

Relations (6-19) and (6-21) can be used for the computation of $X(e^{-j\varphi})$ for any φ. For computation of the discrete Fourier transform it is sufficient to substitute

$$\varphi = k \,.\, 2\pi/N.$$

The relation (6-17) can be interpreted as a linear discrete system into which a sequence x_i enters as an input signal. The value of its response

$z_{N-1}\,\mathrm{e}^{-\mathrm{j}(N-1)\varphi}$ gives the quantity $X(\mathrm{e}^{-\mathrm{j}\varphi})$; in the case $\varphi = k2\pi/N$ the response expresses the discrete Fourier transform

$$X_k = \mathrm{e}^{\mathrm{j}k2\pi/N} \cdot z_{N-1}.$$

Relations (6-19) were initially derived by Goertzel [44] and are known as the Goertzel algorithm [45], [66]. These relations can be further modified [32], [33].

B. Bluestein algorithm

Another way of computing the discrete Fourier transform is given by the Bluestein algorithm [14], converting again the discrete Fourier transform computation into a computation of a convolution. By substituting

$$\text{(6-23)} \qquad ik = \tfrac{1}{2}[i^2 + k^2 - (k - i)^2]$$

we get for the sequence x_i, $i = 0, 1, \ldots, N - 1$, from the defining relation

$$\text{(6-24)} \qquad X_k = \sum_{i=0}^{N-1} x_i W^{ik} = W^{-k^2/2} \sum_{i=0}^{N-1} (x_i W^{-i^2/2})\, W^{(k-i)^2/2},$$

$$k = 0, 1, \ldots, N - 1.$$

This relation contains a discrete convolution of the input sequence x_i multiplied by $W^{-i^2/2} = \mathrm{e}^{-\mathrm{j}i^2\pi/N}$. The result of the convolution is also modified by a multiplication by $W^{-k^2/2} = \mathrm{e}^{-\mathrm{j}k^2\pi/N}$. The algorithm is applicable for any N. The computation of a convolution that is not periodic can be made e.g. by means of FFT for $N' > N$, $N' = 2^m$, padding it with zero terms as is shown in § 7.1.3a.

The algorithm described can be used in signal processing [71] and is employed practically in a device for spectral analysis [18].

6.5 Algorithms for economical computation of the discrete Fourier transform

The algorithm of FFT is valid for complex x_i. In the case where we process real sequences it is not, therefore, fully used. This permits us to make some computational modifications, so that the full capabilities of the computation would be realized [24].

6.5.1 An algorithm for parallel computing of the discrete Fourier transform of two real sequences

(a) A direct transform

Given two real sequences x_i and y_i, $i = 0, 1, \ldots, N - 1$, let us determine their images X_k and Y_k. We set up a complex sequence $z_i = x_i + \mathrm{j}y_i$ and carry out the computation of its discrete Fourier transform $Z_k = \mathscr{D}_N\{z_i\}$. Using relations (4-40) and (4-41) from § 4.1.2 for determination of images of the real and imaginary parts of the sequence and substituting $Z_{-k} = Z_{N-k}$, we get

$$X_k = \frac{1}{2}(Z_k + Z^*_{N-k}), \tag{6-25}$$

$$Y_k = \frac{1}{2j}(Z_k - Z^*_{N-k}). \tag{6-26}$$

Because of the symmetric properties of X_k and Y_k it is sufficient to perform their computation for $k = 0, 1, \ldots \lceil N/2 \rceil$ (where, again, the symbol $\lceil N/2 \rceil$ denotes the integer part of $N/2$).

(b) An inverse transform

By analogy we can perform a computation in parallel of the inverse transform of two real sequences with images $X_k \triangleq x_i$ and $Y_k \triangleq y_i$. We set up from the specified sequences X_k and Y_k the sequence

$$Z_k = X_k + jY_k, \tag{6-27}$$

which is the image of a sequence $x_i + jy_i = z_i$. By the inverse transform of Z_k we get a complex sequence $z_i = \mathscr{D}_N^{-1}\{Z_k\}$ for whose real and imaginary parts

$$x_i = \frac{1}{2}(z_i + z_i^*), \qquad i = 0, 1, \ldots, N-1, \tag{6-28}$$

$$y_i = \frac{1}{2j}(z_i - z_i^*), \qquad i = 0, 1, \ldots, N-1. \tag{6-29}$$

6.5.2 An algorithm for the computation of the discrete Fourier transform of a sequence from the discrete Fourier transform of its even and odd terms

(a) A direct transform

We split the sequence x_i, generally a complex one, of length N (N even), into a selected sequence of even terms $x_{2\nu} = s_\nu^{\text{even}}$ and a sequence of odd terms $x_{2\nu+1} = s_\nu^{odd}$, $\nu = 0, 1, \ldots, N/2 - 1$. In § 6.2.1 we derived the relation (6-4) giving the discrete Fourier transform of a sequence x_i from a discrete Fourier transform of a sequence of even and odd terms. The relation we now write in the form

$$X_k = S_k^{\text{even}} + e^{-jk2\pi/N} S_k^{\text{odd}}, \qquad k = 0, 1, \ldots, N-1, \tag{6-30}$$

where

$$S_k^{\text{even}} = \mathscr{D}_{N/2}\{s_i^{\text{even}}\} \quad \text{and} \quad S_k^{\text{odd}} = \mathscr{D}_{N/2}\{s_i^{\text{odd}}\}. \tag{6-31}$$

This algorithm is based upon the same principle as the first step of the FFT algorithm on the basis of selected input sequences.

Relation (6-30) may be expressed as two relations

(6-32a) $X_k = S_k^{\text{even}} + e^{-jk2\pi/N} S_k^{\text{odd}}, \quad k = 0, 1, \ldots, N/2 - 1,$

(6-32b) $X_{k+N/2} = S_k^{\text{even}} - e^{-jk2\pi/N} S_k^{\text{odd}}, \quad k = 0, 1, \ldots, N/2 - 1.$

which enable us to determine all X_k.

If the sequence x_i is real, one can use the symmetrical properties of its transform and it suffices to compute X_k, e.g. by (6-32a) for $k = 0, 1, \ldots, N/2$ and to determine the other X_k as complex conjugate values $X_{N-k} = X_k^*$.

In the case where $N/2$ is, in addition, even, the computation of X_k for $k = 0, 1, \ldots, N/2$ can be performed by the relations

(6-33a) $X_k = S_k^{\text{even}} + e^{-jk2\pi/N} S_k^{\text{odd}}, \quad k = 0, 1, \ldots, N/4,$

(6-33b) $X_{N/2-k}^* = S_k^{\text{even}} - e^{-jk2\pi/N} S_k^{\text{odd}}, \quad k = 0, 1, \ldots, N/4.$

(b) An inverse transform

We now mention an economical algorithm for computing an inverse discrete Fourier transform by turning it into the inverse discrete Fourier transform of a sequence of half the length. It follows from relations (6-32a) and (6-32b) that by solving them one can easily determine images of even and of odd terms from the image X_k of the complete sequence†)

(6-34a) $S_k^{\text{even}} = \frac{1}{2}(X_k + X_{N/2+k}), \quad k = 0, 1, \ldots, N/2 - 1,$

(6-34b) $S_k^{\text{odd}} = \frac{1}{2}(X_k - X_{N/2+k})\, e^{jk2\pi/N}, \quad k = 0, 1, \ldots, N/2 - 1.$

By inverse discrete Fourier transforming we then get

$$s_i^{\text{even}} = \mathscr{D}_{N/2}^{-1}\{S_k^{\text{even}}\}, \quad i = 0, 1, \ldots, N/2 - 1,$$

$$s_i^{\text{odd}} = \mathscr{D}_{N/2}^{-1}\{S_k^{\text{odd}}\}, \quad i = 0, 1, \ldots, N/2 - 1,$$

(6-35) $s_i^{\text{even}} = x_{2i} \quad \text{and} \quad s_i^{\text{odd}} = x_{2i+1}.$

If the sequence X_k should be the transform of a real sequence and $N/2$ even, we can begin from (6-33a) and (6-33b) and get by their solution

(6-36a) $S_k^{\text{even}} = \frac{1}{2}(X_k + X_{N/2-k}^*), \quad k = 0, 1, \ldots, N/4,$

(6-36b) $S_k^{\text{odd}} = \frac{1}{2}(X_k - X_{N/2-k}^*)\, e^{jk2\pi/N}, \quad k = 0, 1, \ldots, N/4.$

By conjugating terms S_k^{even} and S_k^{odd} calculated in this manner we may work out the full numbers and we calculate s_i^{even} and s_i^{odd} using $\mathscr{D}_{N/2}^{-1}$.

†) It is obvious that we could obtain the expressions for S_k^{even} and S_k^{odd} from values X_k by the direct computation of § 4.3.3 (see (4-118) and (4-119)).

6.5.3 An algorithm for the economical computation of the discrete Fourier transform of a real sequence

(a) The direct transform

Let us apply first the algorithm from § 6.5.2 and split the sequence x_i into the sequence of even terms s_i^{even} and that of odd terms s_i^{odd}. The transforms of these sequences $s_i^{\text{even}} \triangleq S_k^{\text{even}}$ and $s_i^{\text{odd}} \triangleq S_k^{\text{odd}}$, necessary for computation by (6-32) or by (6-33), are determined by the algorithm from § 6.5.1 for parallel computation of the discrete Fourier transforms.

Thence we shall take the following steps:

1. The sequence $x_i = 0, 1, \ldots, N-1$ ($N/2$ even) we split into even terms $s_i^{\text{even}} = x_{2i}$ and odd terms $s_i^{\text{odd}} = x_{2i+1}$, $i = 0, 1, \ldots, N/2-1$ and we create a complex sequence $z_i = s_i^{\text{even}} + \mathrm{j}s_i^{\text{odd}}$.

2. We perform the discrete Fourier transform of the sequence z_i and obtain

$$Z_k = \mathscr{D}_{N/2}\{z_i\}.$$

3. From the sequence Z_k we determine by (6-25) and (6-26) the transforms $S_k^{\text{even}} \triangleq s_i^{\text{even}}$ and $S_k^{\text{odd}} \triangleq s_i^{\text{odd}}$

$$S_k^{\text{even}} = \frac{1}{2}(Z_k + Z^*_{N/2-k}),$$

$$S_k^{\text{odd}} = \frac{1}{2\mathrm{j}}(Z_k - Z^*_{N/2-k}).$$

4. From the transforms S_k^{even} and S_k^{odd} we determine by (6-33) X_k for $k = 0, 1, \ldots, N/2$:

$$X_k = S_k^{\text{even}} + \mathrm{e}^{-\mathrm{j}k2\pi/N} S_k^{\text{odd}}, \qquad k = 0, 1, \ldots, N/4-1,$$
$$X^*_{N/2-k} = S_k^{\text{even}} - \mathrm{e}^{-\mathrm{j}k2\pi/N} S_k^{\text{odd}}, \qquad k = 0, 1, \ldots, N/4.$$

5. By conjugating we complete the X_k to N values.

(b) The inverse transform

In this case we proceed so that to a given $X_k \triangleq x_i$ we determine by the algorithm (§ 6.5.2b) transforms of sequences of even and odd terms. Then by the algorithm from § 6.5.1b we determine the complex sequence corresponding to them by (6-27). The whole procedure may be divided into these steps:

1. From the sequence X_k, $k = 0, 1, \ldots, N-1$, we compute S_k^{even} and S_k^{odd} by (6-36):

$$S_k^{\text{even}} = \tfrac{1}{2}(X_k + X^*_{N/2-k}), \qquad k = 0, 1, \ldots, N/4,$$
$$S_k^{\text{odd}} = \tfrac{1}{2}(X_k - X^*_{N/2-k})\,\mathrm{e}^{\mathrm{j}k2\pi/N}, \qquad k = 0, 1, \ldots, N/4,$$

and complete them by conjugating to $k = 0, 1, \ldots, N/2-1$.

2. We set up the complex sequence

$$Z_k = S_k^{\text{even}} + \mathrm{j}S_k^{\text{odd}}, \qquad k = 0, 1, \ldots, N/2 - 1.$$

3. We perform the discrete Fourier transform

$$\mathscr{D}_{N/2}^{-1}\{Z_k\} = z_i = s_i^{\text{even}} + \mathrm{j}s_i^{\text{odd}}, \qquad i = 0, 1, \ldots, N/2 - 1.$$

4. From the sequences s_i^{even} and s_i^{odd} we determine x_i:

$$x_{2i} = s_i^{\text{even}}, \qquad i = 0, 1, \ldots, N/2 - 1,$$
$$x_{2i+1} = s_i^{\text{odd}}, \qquad i = 0, 1, \ldots, N/2 - 1.$$

7. SOME APPLICATIONS OF DISCRETE FOURIER TRANSFORMS

Most applications of discrete Fourier transforms lie in the area of computing Fourier transforms that we have been concerned with up till now. Classical methods of spectral analysis are linked to this application. During almost ten years of use of the discrete Fourier transform a large number of new applications have appeared, some of them very special, such as so-called deconvolution [79], homomorphic filtering [66], short time spectral analysis [72], Fourier spectroscopy and the like. We present here one of the most common examples of an application of discrete Fourier transforms, i.e. a fast computation of a convolution. For special cases we present in Chapter 8 an application of discrete Fourier transforms to the computation of the Hilbert transform.

7.1 Application of the discrete Fourier transform to the computation of a convolution

7.1.1 The difference between the convolution and the periodic convolution

In § 4.2.2 we have shown that the discrete periodic convolution corresponds to a product of image sequences. We have assumed in that case that both sequences are periodic with the same period. In practice and especially in the theory of linear discrete systems, however, a discrete non-periodic convolution occurs, of the form

$$\text{(7-1)} \qquad \sum_{\nu=-\infty}^{\infty} x_\nu y_{i-\nu} = \sum_{\nu=-\infty}^{\infty} x_{i-\nu} y_\nu = x_i * y_i, \qquad -\infty < i < \infty.$$

We shall call this convolution *the discrete convolution* or just the convolution. If a sequence x_i, $i = 0, 1, \ldots, N_1 - 1$ has a number of terms N_1 and a sequence y_i, $i = 0, 1, \ldots, N_2 - 1$, a number of terms N_2, then the convolution can be written in the form

$$\text{(7-2)} \qquad x_i * y_i = \sum_{\nu=0}^{i} x_\nu y_{i-\nu} = \sum_{\nu=0}^{i} y_\nu x_{i-\nu},$$
$$i = 0, 1, \ldots, N-1, \quad N = N_1 + N_2 - 1,$$

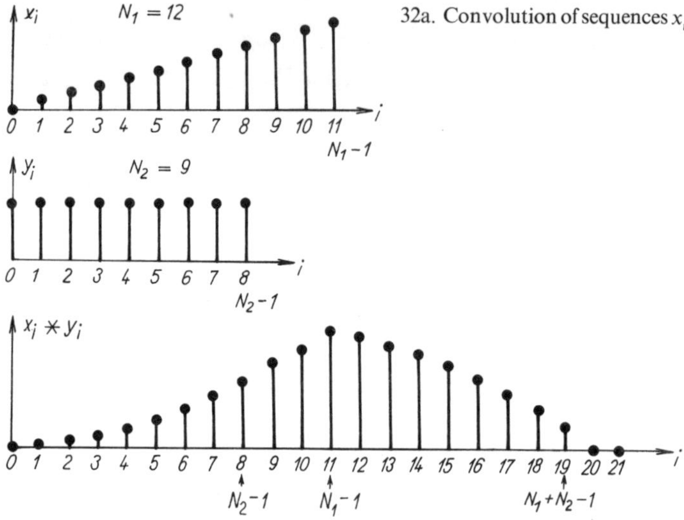

32a. Convolution of sequences x_i and y_i

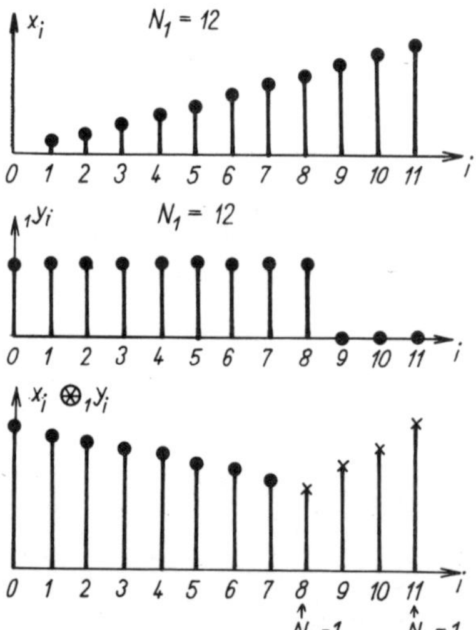

32b. Periodic convolution of a sequence x_i with a padded sequence ${}_1y_i$

and defined only in the specified interval. The convolution of two finite sequences of lengths N_1 and N_2 therefore has a length $N_1 + N_2 - 1$. In Figure 32(a) the convolution of two sequences of different lengths is shown. In Figure 32(b) is shown the periodic convolution of these sequences, where the sequence y_i is padded with zero terms to the sequence ${}_1y_i$ of the same length as the sequence x_i, i.e. N_1, $N_1 > N_2$, which enables us to perform the periodic convolution. Between the convolution and the periodic convolution the following relations hold:

1. For the convolution of sequences x_i, $i = 0, 1, \ldots, N - 1$, and y_i, $i = 0, 1, \ldots, N_2 - 1$, $N_1 > N_2$, only terms with indices $i = N_2 - 1$,

$N_2, \ldots, N_1 - 1$ can be determined by the periodic convolution of the sequence x_i with the padded sequence ${}_1y_i$

$$(7\text{-}3) \qquad {}_1y_i = \begin{cases} y_i, & i = 0, 1, \ldots, N_2 - 1 \\ 0, & i = N_2, N_2 + 1, \ldots, N_1 - 1. \end{cases}$$

These terms are marked differently in Figure 32(b).

2. The first $N_2 - 1$ terms of the periodic convolution do not identify with the corresponding terms of the convolution.

3. The last N_2 terms of the convolution, i.e. the terms with indices satisfying the inequality $N_1 \leqq i \leqq N_1 + N_2 - 1$, cannot be determined by means of the periodic convolution.

It therefore follows that in the case $N_2 = N_1$ it is not possible, by the procedure described, using the periodic convolution, to determine any term of the convolution but the term with index $N_1 - 1$.

7.1.2 Computation of a convolution by means of a periodic convolution

If we want to determine all terms of a convolution by means of a periodic convolution, we must modify the sequence x_i of length N_1 and the sequence y_i of length N_2 so that at their periodic lengthening an interaction occurs only between one period of each of them. For that one must carry out the following set-up.

1. Choose $N \geqq N_1 + N_2 - 1$;
2. Complete the sequence x_i to ${}_1x_i$:

$${}_1x_i = \begin{cases} x_i, & i = 0, 1, \ldots, N_1 - 1, \\ 0, & i = N_1, N_1 + 1, \ldots, N - 1; \end{cases}$$

3. Complete the sequence y_i to ${}_1y_i$:

$${}_1y_i = \begin{cases} y_i, & i = 0, 1, \ldots, N_2 - 1, \\ 0, & i = N_2, N_2 + 1, \ldots, N - 1. \end{cases}$$

In the convolution computation itself we use the discrete Fourier transform to perform, according to theorem 13 (Chapter 4), the following computations

1. ${}_1X_k = \mathscr{D}_N\{{}_1x_i\}$,
2. ${}_2X_k = \mathscr{D}_N\{{}_1y_i\}$,
3. ${}_1X_k\,{}_2X_k = Z_k$,
4. $\mathscr{D}_N^{-1}\{Z_k\} = z_i$.

The first $N_1 + N_2 - 1$ terms of the sequence z_i then give the convolution values required.

Though this procedure may seem tedious, it is faster for complex sequences for $N > 64$ when using the base-2 FFT algorithm than a direct computation by the defining relation for the convolution [17]. The computation may be made yet faster in several ways, e.g. for real sequences it is possible to compute in parallel by the algorithm of § 6.5.1a the transform of sequences ${}_1x_i$ and ${}_1y_i$ and an inverse transform can be performed by the algorithm from § 6.5.3b. The way described for computing the convolution is usually called in the literature *the fast convolution.*

Recently several methods for the direct computation of a convolution were worked out, which require only a very small number of multiplications and do not use the discrete Fourier transform. Their theoretical basis is founded upon analogous considerations to those indicated in § 6.4.2. More detailed information about these methods can be found [2], [65] and [61].

Another group of convolution computation methods is represented by the number theoretic transforms (NTT) [61], which work with modular arithmetic and eliminate round-off errors; they have, however, also some drawbacks.

7.1.3 Computation of the convolution of a long sequence

If one of the sequences whose convolution we intend to compute is very long, so that it cannot, for example, be placed in the memory of a computer as a unit, the computation must be split into sections. We shall

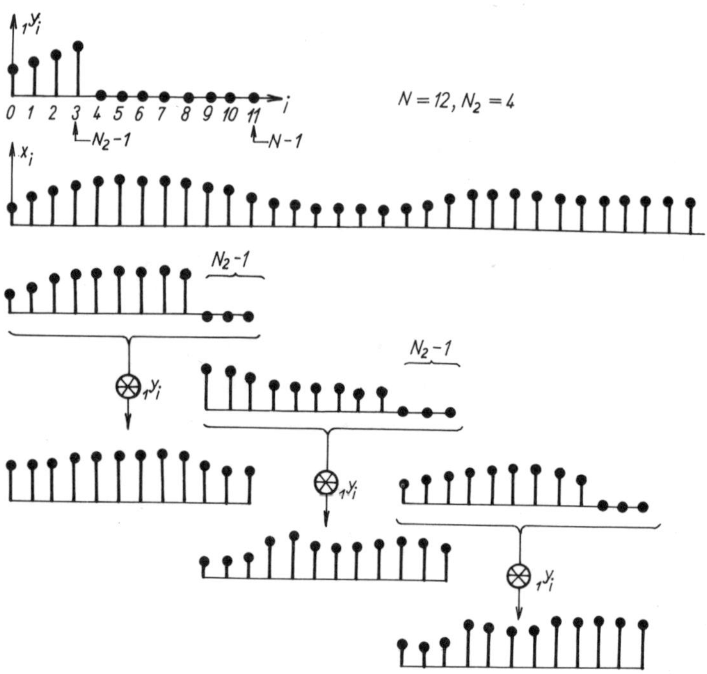

33. Procedure for computation of a convolution with a long sequence by the overlap-add algorithm

assume that N_2 is the length of the shorter sequence and that we have chosen a suitable $N > N_2$. Here we describe and illustrate graphically two procedures:

(A) An algorithm for the computation of the convolution called the overlap-add algorithm (see Figure 33):

1. We split the sequence x_i into sections of length $N_1 = N - (N_2 - 1)$.
2. From each section we set up a partial sequence of length N, so that we pad it with $N_2 - 1$ zero terms.
3. We perform successively the computation of the convolution of each of the so created partial sequences with the sequence ${}_1y_i$ by the algorithm from § 7.1.2. The result will again be sequences of length N.
4. The sequences of length N obtained in this way are sorted sequentially so that each following sequence overlaps the last $N_2 - 1$ terms of the preceding sequence. The resulting sequence is set up from terms of the single sequences sorted in the way mentioned and, where the neighbouring sequences overlap, from the sums of the overlapping terms.

(B) An algorithm for the computation of the convolution called the overlap-save algorithm (see Figure 34).

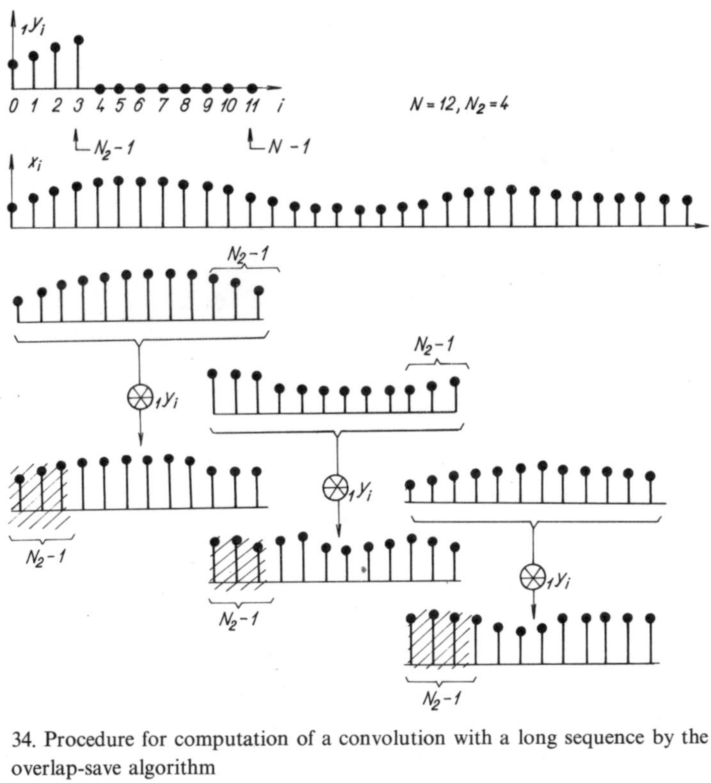

34. Procedure for computation of a convolution with a long sequence by the overlap-save algorithm

1. We split the sequence x_i into partial sequences of length N, which overlap so that each following sequence contains as its first terms the last $N_2 - 1$ terms of the preceding sequence.

2. We perform a computation of the periodic convolution of the single partial sequences with the sequence ${}_1y_i$ and so obtain again a sequence of length N.†)

3. In each such sequence we drop the first $N_2 - 1$ terms.

4. The sequences adjusted in this manner are sorted sequentially. Thus the resulting sequence arises.

Both methods are roughly equally fast. E. O. Brigham [17] recommends the use of the optimum values of the length of partial sequences N depending on the length of the short sequence according to Table 6.

Table 6

N_2	N	N_2	N
11–19	64	200–299	2 048
20–29	128	300–599	4 096
30–49	256	600–999	8 192
50–99	512	1 000–1 999	16 384
100–199	1 024	2 000–3 999	32 768

7.2 Application of discrete Fourier transforms to computations of correlation and of power spectrum density

7.2.1 Relation of the periodic correlation to the correlation

In § 4.2.3 we defined the discrete periodic correlation by relation (4-70)

$$\varrho_i(x_\nu, y_\nu) = \sum_{\nu=0}^{N-1} x_\nu y_{\nu+i}$$

to which there applies:

$$\varrho_i(x_\nu, y_\nu) = X_{-k} Y_k .$$

We now demonstrate a computation of the correlation sequence

$$R_i(x_\nu, y_\nu) = \sum_{\nu=0}^{N_1-1} x_\nu y_{\nu+i}, \qquad -(N_1 - 1) \leqq i \leqq N_2 - 1, \tag{7-4}$$

†) In this case we cannot directly use the algorithm from § 7.1.2, as the partial sequence of length N taken from the given sequence has not been padded with zero terms, but we perform directly its periodic convolution with the sequence ${}_1y_i$. The terms by which the computed periodic convolution differs from the convolution are then dropped.

defined for finite real sequences x_ν, $\nu = 0, 1, \ldots, N_1 - 1$, and y_ν, $\nu = 0, 1, \ldots, N_2 - 1$. In order to use the periodic correlation for the computation of $R_i(x_\nu, y_\nu)$ and hence the discrete Fourier transform, we must appreciate the effect of periodized sequences which arises with the discrete Fourier transform by analogy with the convolution computation. We use the following procedure.

1. We set up padded sequences of length $N \geqq N_1 + N_2 - 1$:

$$_1x_\nu = \begin{cases} x_\nu, & \nu = 0, 1, \ldots, N_1 - 1, \\ 0, & \nu = N_1, N_1 + 1, \ldots, N - 1, \end{cases}$$

$$_2y_\nu = \begin{cases} 0, & \nu = 0, 1, \ldots, N - N_2 - 1, \\ y_{\nu-(N-N_2)}, & \nu = N - N_2, \ldots, N - 1. \end{cases}$$

2. We compute the transforms

$$\mathscr{D}_N\{_1x_\nu\} = {_1X_k},$$
$$\mathscr{D}_N\{_2y_\nu\} = {_2Y_k}.$$

3. By (4-74) we compute

$$_2Z_k = {_1X_k}\,{_2Y_{-k}}.$$

4. We perform the inverse transform

$$\mathscr{D}_N^{-1}\{_2Z_k\} = \varrho_i(x_\nu, y_\nu).$$

5. We determine

$$R_i(x_\nu, y_\nu) = \varrho_{i+N-N_2}(x_\nu, y_\nu), \qquad -(N_1 - 1) \leqq i \leqq N_2 - 1.$$

If we want to use this procedure to determine the autocorrelation sequence

$$R_i(x_\nu, x_\nu) = \sum_{\nu=0}^{N_1-1} x_\nu x_{\nu+i}, \qquad -(N_1 - 1) \leqq i \leqq N_1 - 1,$$

of the sequence x_ν, $\nu = 0, 1, \ldots, N_1 - 1$, then from the relations presented for the correlation the following procedure results:

1. We set up a sequence

$$_1x_\nu = \begin{cases} x_\nu, & \nu = 0, 1, \ldots, N_1 - 1, \\ 0, & \nu = N_1, \ldots, N - 1. \end{cases}$$

2. We compute the transform

$$\mathscr{D}_N\{_1x_\nu\} = {_1X_k}.$$

3. We set up by (4-78) the sequence

$$_2Z_k = |{_1X_k}|^2.$$

4. We perform the inverse transform

$$\mathscr{D}_N^{-1}\{_2Z_k\} = \varrho_i(x_\nu, x_\nu).$$

5. We determine

$$R_i(x_v, x_v) = \begin{cases} \varrho_i(x_v, x_v), & i = 0, 1, \ldots, N_1 - 1, \\ \varrho_{-i}(x_v, x_v), & i = -1, \ldots, -(N_1 - 1). \end{cases}$$

7.2.2 Computation of power spectrum density

The computation of power spectrum density, or more precisely its estimation, presents an independent problem in which the way of using the discrete Fourier transform is controlled by a chosen method of computation. Standard methods are described e.g. in [71], [53], [5], [60], [13], [1], [11].

7.3 Two-dimensional discrete Fourier transforms

Two-dimensional discrete Fourier transforms are defined by the relation

$$(7\text{-}5) \qquad X_{k,l} = \sum_{m=0}^{M-1} \sum_{n=0}^{N-1} x_{m,n}\, e^{-jmk2\pi/M}\, e^{-jml2\pi/N},$$

where $x_{m,n}$ is a two-dimensional sequence, $m = 0, 1, \ldots, M - 1$, $n = 0, 1, \ldots, N - 1$, that is assumed to be periodic with a period M in an index m and with a period N in an index n.

The inverse discrete Fourier transform is then expressed by the relation

$$(7\text{-}6) \qquad x_{m,n} = (MN)^{-1} \sum_{k=0}^{M-1} \sum_{l=0}^{N-1} X_{k,l}\, e^{jmk2\pi/M}\, e^{jnl2\pi/N}.$$

When comparing relation (7-5) with relation (4-121) from § 4.3.4 where we modified a one-dimensional sequence by reindexing into the two-dimensional one, there is an obvious difference in that the definition of the two-dimensional (2-*D*) discrete Fourier transform does not contain in its kernel a product of indices *ml* or *nk*. As a result it is possible to turn the 2-*D* discrete Fourier transform into the successive performing of 1-*D* discrete Fourier transforms either in this order

$$(7\text{-}7a) \qquad {}_nX_k = \sum_{m=0}^{M-1} x_{m,n}\, e^{-jmk2\pi/M},$$

$$k = 0, 1, \ldots, M - 1, \quad n = 0, 1, \ldots, N - 1,$$

$$(7\text{-}7b) \qquad X_{k,l} = \sum_{n=0}^{N-1} {}_nX_k\, e^{-jnl2\pi/N},$$

$$l = 0, 1, \ldots, N - 1, \quad k = 0, 1, \ldots, M - 1,$$

or in reverse order

$$(7\text{-}8a)\quad {}_mX_l = \sum_{n=0}^{N-1} x_{m,n}\, \mathrm{e}^{-\mathrm{j}nl2\pi/N},$$

$$l = 0, 1, \ldots, N-1, \quad m = 0, 1, \ldots, M-1,$$

$$(7\text{-}8b)\quad X_{k,l} = \sum_{m=0}^{M-1} {}_mX_l\, \mathrm{e}^{-\mathrm{j}mk2\pi/N},$$

$$k = 0, 1, \ldots, M-1, \quad l = 0, 1, \ldots, N-1.$$

The computation of the 2-*D* discrete Fourier transforms thus requires N computations of the transform $\mathscr{D}_M$ and M computations of the transform $\mathscr{D}_N$. If we imagine $x_{m,n}$ as a matrix, then we carry out $\mathscr{D}_N$ first by rows. The N sequences created can again be set up into a matrix in which we carry out successively the computation of $\mathscr{D}_N$ from single columns. By analogy the computation of the inverse transform can also be expanded.

The 2-*D* discrete Fourier transform is of great importance for image processing [4], [7].

8. DISCRETE HILBERT TRANSFORMS

Hilbert transforms have many applications in electrical engineering, optics and quantum physics. We show here, therefore, some possibilities of using the discrete Fourier transform for their calculation.

8.1 Computation of Hilbert transforms by means of the discrete Fourier transform

If we are to determine the Hilbert transform of a periodic function x given just by its values $x(iT)$, we may proceed by constructing through the values x_i a trigonometric polynomial by § 2.2.2 and determine its Hilbert transform. This procedure is analogous to the construction of a conjugate Fourier series, as presented in § 2.4.5.

We defined the Hilbert transform in § 2.4.5 by the relation

$$y(t) = \frac{1}{\pi}\int_{-\infty}^{\infty} \frac{x(\tau)}{\tau - t}\, d\tau = \mathscr{H}\{x(t)\}, \tag{8-1}$$

which can be written also as a convolution

$$y(t) = -\frac{1}{\pi t} * x(t). \tag{8-2}$$

For an approximation to a function x defined in points $x(iT) = x_i$, $i = 0, 1, \ldots, N-1$, $T = P/N$ by a trigonometric polynomial u_M,

$$u_M(t) = \sum_{k=-M}^{M} D_k\, e^{jk2\pi t/P}, \tag{8-3}$$

we derived in § 2.2.1 the relations for coefficients D_k and in § 5.1.3 we showed that the coefficients D_k may be expressed from values of the function x_i, $i = 0, 1, \ldots, N-1$, using their discrete Fourier transform X_k. Then the polynomial u_M can be written out for $M \leqq (N-1)/2$ in the form

$$u_M(t) = \frac{1}{N} X_0 + \frac{1}{N}\sum_{k=1}^{M} (X_k\, e^{-jk2\pi t/P} + X_{N-k}\, e^{-jk2\pi t/P}) \tag{8-4}$$

or in the form

$$(8\text{-}5)\qquad u_M(t) = \frac{1}{N}X_0 + \frac{1}{N}\sum_{k=1}^{M-1}(X_k\,\mathrm{e}^{\mathrm{j}k2\pi t/P} + X_{N-k}\,\mathrm{e}^{-\mathrm{j}k2\pi t/P}) + \\ + \frac{1}{2N}X_{N/2}(\mathrm{e}^{\mathrm{j}k2\pi t/P} + \mathrm{e}^{-\mathrm{j}k2\pi t/P})$$

for $M = N/2$.

Let us now compute the Hilbert transform of the function u_M. We know that (2-72) holds, i.e.

$$\mathscr{H}\{\mathrm{e}^{\mathrm{j}\omega_0 t}\} = \mathrm{e}^{\mathrm{j}\omega_0 t}\,\mathrm{j}\,.\,\mathrm{sign}\,\omega_0\,;$$

if we use

$$(8\text{-}6)\qquad v_M(t) = \mathscr{H}\{u_M(t)\},$$

we get for $M \leqq (N-1)/2$

$$(8\text{-}7)\qquad v_M(t) = \frac{1}{N}\sum_{k=1}^{M}(\mathrm{j}X_k\,\mathrm{e}^{\mathrm{j}k2\pi t} - \mathrm{j}X_{N-k}\,\mathrm{e}^{-\mathrm{j}k2\pi t/P})$$

and for $M = N/2$

$$(8\text{-}8)\qquad v_M(t) = \frac{1}{N}\sum_{k=1}^{M-1}(\mathrm{j}X_k\,\mathrm{e}^{\mathrm{j}k2\pi t/P} - \mathrm{j}X_{N-k}\,\mathrm{e}^{-\mathrm{j}k2\pi t/P}) + \\ + \frac{1}{2N}\mathrm{j}X_{N/2}(\mathrm{e}^{\mathrm{j}N\pi t/P} - \mathrm{e}^{-\mathrm{j}N\pi t/P}).$$

If we now consider the boundary cases $M = N/2$ and $M = (N-1)/2$ and substitute X_k from the defining relation of the discrete Fourier transform, we can write the Hilbert transform of the function u in the form of trigonometric interpolation polynomials

$$(8\text{-}9)\qquad v(t) = \begin{cases} -\dfrac{1}{N}\displaystyle\sum_{n=0}^{N-1} x_n\left[1 - \dfrac{\cos\pi(t-nT)/T}{\cos\pi(t-nT)\,NT}\right]\mathrm{cotg}\,\pi\dfrac{t-nT}{NT}, & N \text{ odd}, \\ -\dfrac{1}{N}\displaystyle\sum_{n=0}^{N-1} x_n[1 - \cos\pi(t-nT)/T]\,\mathrm{cotg}\,\pi\dfrac{t-nT}{NT}, & N \text{ even}. \end{cases}$$

If we substitute in (8-7) and (8-8) $t = iT$, they turn into the relations

$$(8\text{-}10)\qquad v_M(iT) = \frac{1}{N}\sum_{k=1}^{M}(\mathrm{j}X_k\,\mathrm{e}^{\mathrm{j}ik2\pi/N} - \mathrm{j}X_{N-k}\,\mathrm{e}^{-\mathrm{j}ik2\pi/N}), \qquad M \leqq (N-1)/2,$$

$$(8\text{-}11)\qquad v_M(iT) = \frac{1}{N}\sum_{k=1}^{M-1}(\mathrm{j}X_k\,\mathrm{e}^{\mathrm{j}ik2\pi/N} - \mathrm{j}X_{N-k}\,\mathrm{e}^{-\mathrm{j}ik2\pi/N}), \qquad M = N/2.$$

If we consider the boundary cases $M = (N-1)/2$ and $M = N/2$ and distinguish only even and odd N, we get from (8-10) and (8-11) an expression for $v(iT)$:

$$v(iT) = \begin{cases} \dfrac{1}{N} \sum_{k=1}^{(N-1)/2} (\mathrm{j}X_k \, \mathrm{e}^{\mathrm{j}ik2\pi/N} - \mathrm{j}X_{N-k} \, \mathrm{e}^{-\mathrm{j}ik2\pi/N}), & N \text{ odd}, \\ \\ \dfrac{1}{N} \sum_{k=1}^{N/2-1} (\mathrm{j}X_k \, \mathrm{e}^{\mathrm{j}ik2\pi/N} - \mathrm{j}X_{N-k} \, \mathrm{e}^{-\mathrm{j}ik2\pi/N}), & N \text{ even}, \end{cases} \tag{8-12}$$

that can be rewritten to a form corresponding to the discrete Fourier transform

$$v(iT) = \begin{cases} \dfrac{1}{N} \left(\sum_{k=1}^{(N-1)/2} \mathrm{j}X_k \, \mathrm{e}^{\mathrm{j}ik2\pi/N} - \sum_{(N+1)/2}^{N-1} \mathrm{j}X_k \, \mathrm{e}^{\mathrm{j}ik2\pi/N} \right), & N \text{ odd}, \\ \\ \dfrac{1}{N} \left(\sum_{k=1}^{N/2-1} \mathrm{j}X_k \, \mathrm{e}^{\mathrm{j}ik2\pi/N} - \sum_{N/2+1}^{N-1} \mathrm{j}X_k \, \mathrm{e}^{\mathrm{j}ik2\pi/N} \right), & N \text{ even}. \end{cases} \tag{8-13}$$

If we now introduce

$$Y_k = \begin{cases} \mathrm{j}X_k, & k = 1, 2, \ldots, (N-1)/2, \\ 0, & k = 0, \\ -\mathrm{j}X_k, & k = (N+1)/2, \ldots, N-1, \end{cases} \qquad N \text{ odd}, \tag{8-14}$$

$$Y_k = \begin{cases} \mathrm{j}X_k, & k = 1, 2, \ldots, N/2-1, \\ 0, & k = 0, N/2, \\ -\mathrm{j}X_k, & k = N/2+1, \ldots, N-1, \end{cases} \qquad N \text{ even}, \tag{8-15}$$

and put $v(iT) = y_i$, then it is obvious that (8-13) may be expressed as an inverse Fourier transform

$$y_i = \frac{1}{N} \sum_{k=0}^{N-1} Y_k \, \mathrm{e}^{\mathrm{j}ik2\pi/N}. \tag{8-16}$$

If we also introduce

$$H_k = \begin{cases} \mathrm{j}, & k = 1, 2, \ldots, (N-1)/2, \\ 0, & k = 0, \\ -\mathrm{j}, & k = (N+1)/2, \ldots, N-1, \end{cases} \qquad N \text{ odd}, \tag{8-17}$$

and

$$H_k = \begin{cases} \mathrm{j}, & k = 1, 2, \ldots, N/2-1, \\ 0, & k = 0, N/2, \\ -\mathrm{j}, & k = N/2+1, \ldots, N-1, \end{cases} \qquad N \text{ even}, \tag{8-18}$$

then we can write Y_k as a product

(8-19) $\quad Y_k = X_k H_k.$

This expression is analogous to the relation (2-71) which expresses the Fourier transform in terms of the Hilbert transform. It follows from (8-19) that we can compute the Hilbert transform y_i of a periodic sequence x_i by means of the discrete Fourier transform by the following procedure using theorem 13a on discrete periodic convolution:

1. We carry out the calculation of $X_k = \mathscr{D}_n\{x_i\}$.
2. We compute $Y_k = X_k H_k$.
3. We carry out an inverse discrete Fourier transform

$$y_i = \mathscr{D}_N^{-1}\{Y_k\}.$$

This procedure is accurate in the cases where the X_k correspond to actual values of the Fourier transform of a function x whose discrete values we know. It follows from § 5.1.1 that this is true for all periodic functions whose Fourier series coefficients C_k are equal to zero (or at least are negligible) for $k \geqq N/2$.

8.2 Discrete Hilbert transforms

It follows from (8-19) according to theorem 13 of § 4.2.2 that the sequence y_i may be expressed by a periodic convolution

(8-20) $\quad y_i = x_i \circledast h_i, \quad \text{where} \quad h_i = \mathscr{D}^{-1}\{H_k\}.$

If we compute the inverse discrete Fourier transform of images (8-17) and (8-18)

(8-21) $$h_i = -\frac{2}{N}\sum_{k=1}^{L} \sin(ik\,.\,2\pi/N), \qquad i = 0, 1, \ldots, N-1$$

where $L = N/2$ for even N and $L = (N-1)/2$ for odd N, we get the sequence

(8-22) $$h_i = \begin{cases} -\dfrac{1}{N}\left[1 - \dfrac{(-1)^i}{\cos i\pi/N}\right]\operatorname{cotg} i\pi/N, & N \text{ odd}, \\ & i = 1, \ldots, N-1 \\ -\dfrac{1}{N}\left[1-(-1)^i\right]\operatorname{cotg} i\pi N, & N \text{ even}, \end{cases}$$

$$h_0 = 0,$$

which we call *the kernel of a discrete Hilbert transform.* The corresponding periodic convolution

(8-23) $$y_i = \sum_{\nu=0}^{N-1} x_\nu h_{i-\nu}, \qquad i = 0, 1, \ldots, N-1$$

we call *the discrete Hilbert transform* [28], [27]

$$y_i = \mathscr{H}_D \{x_i\}.$$

In the case of even N the transform can be written in the following way:

$$y_i = \begin{cases} -\dfrac{1}{N} \sum\limits_{v=0}^{N/2-1} x_{2v} \operatorname{cotg}(i - 2v)\dfrac{\pi}{N}, & i \text{ odd}, \\ -\dfrac{1}{N} \sum\limits_{v=0}^{(N-1)/2} x_{2v+1} \operatorname{cotg}(i - (2v + 1))\dfrac{\pi}{N}, & i \text{ even}. \end{cases}$$

8.3 Some applications of discrete Hilbert transforms

a) Discrete Hilbert transforms can be used for approximate calculations of Hilbert transforms. In the case of periodic functions the procedure is obvious. In the case of finite aperiodic sequences the calculation is complicated by the fact that the Hilbert transform of a finite function is not finite. It is, therefore, necessary to complete finite sequences by zero elements up to some multiple of their length. It is, however, possible to use a change of variable $t = \operatorname{tg}(s/2)$ [29].

b) Furthermore, we show how the discrete Hilbert transform can be used for the calculation of a so-called discrete analytic signal. The analytic signal corresponding to a real signal is the complex signal whose real part is equal to the given signal and whose imaginary part is equal to its negatively taken Hilbert transform. If the real signal is described by the sequence x_i, $i = 0, 1, \ldots, N - 1$, then the analytic signal is expressed by the sequence $z_i = x_i - \mathrm{j}y_i$; then for its discrete Fourier transform we have $Z_k = X_k - \mathrm{j}Y_k$. If we replace Y_k by $X_k H_k$ and express H_k according to (8-17) and (8-18), we get for even N

$$(8\text{-}24) \qquad Z_k = \begin{matrix} X_k - \mathrm{j}(\mathrm{j}X_k) = 2X_k, & k = 1, 2, \ldots, N/2 - 1, \\ X_k, & k = 0, N/2, \\ X_k - \mathrm{j}(-\mathrm{j}X_k) = 0, & k = 0, N/2, N/2 + 1, \ldots, N - 1. \end{matrix}$$

Thus it is sufficient to carry out the calculation of $\mathscr{D}_N\{x_i\} = X_k$ and then to construct the sequence Z_k by (8-24). If we construct the inverse discrete Fourier transform of X, we get $\mathscr{D}_N^{-1}\{Z_k\} = x_i - \mathrm{j}y_i$. The real part of the resulting sequence is the original signal and the imaginary part is its discrete Hilbert transform.

c) The discrete Hilbert transform can be used for calculating the envelope and instantaneous frequency of a signal [31]. In addition, it is also necessary to determine by means of the discrete Fourier transform the derivative of the function and the derivative of the Hilbert transform. By analogy with the

Table 7.

D_k	K_k	k	
		N even	N odd
jk	$-k$	$1, 2, \ldots, N/2 - 1$	$1, 2, \ldots, (N-1)/2$
0	0	$0, N/2$	0
$j(k - N)$	$k - N$	$N/2 + 1, \ldots, N - 1$	$(N+1)/2, \ldots, N - 1$

way we derived the Hilbert transform from a trigonometric polynomial, a discrete expression for the derivative and the derivative of the Hilbert transform can be derived. If x_i is a given sequence with X_k, then for the derivative transform

$$x'_i \triangleq D_k X_k,$$

$$y'_i \triangleq K_k X_k,$$

where the sequences D_k and K_k are shown in Table 7. The discrete instantaneous frequency ω_i at instants iT is then determined by the sequence

(8-25) $\quad \omega_i = E_i^{-2}(x'_i y_i - x_i y'_i)$

and values of the envelope E_i are expressed by the sequence

(8-26) $\quad E_i = \sqrt{(x_i^2 + y_i^2)}.$

BIBLIOGRAPHY

1. Achilles, D.: Die Fourier-Transformation in der Signalverarbeitung. Berlin, Springer Verlag 1978.
2. Agarwal, R. O.—Cooley, J. W.: New Algorithms for Digital Convolution. *Trans. IEEE,* ASSP-**25**, 1977, no *5*, pp. 392—410.
3. Ahlberg, J. H.—Nilson, E. N.—Walsh, J. L.: The Theory of Splines and Their Applications. New York, Academic Press 1967.
4. Ahmed, N.—Rao, K. R.: Orthogonal Transforms for Digital Signal Processing. Berlin, Springer Verlag 1975.
5. Anděl, J.: Statistická analýza časových řad (Statistical analysis of time series). Prague, SNTL 1976.
6. Angot, A.: Compléments de mathématique à l'usage des ingenieurs de électrotechnique et des télécommunications. Paris, Edition de la Revue d'optique, 1965.
7. Antosik, P.—Mikusinski, J.—Sikorski, R.: Theory of Distribution. Amsterdam, Elsevier Scientific Publishing Comp. 1973.
8. Aplikovaná matematika I. a II. (Applied Mathematics I and II). Prague, SNTL 1977—1979.
9. Arsac, J.: Fourier Transforms and the Theory of Distributions. London, Prentice Hall Inc. 1966.
10. Bauer, F. L.—Stetter, H. J.: Zur numerischen Fourier-Transformation. *Num. Math.* vol. **1**, 1959, pp. 208—220.
11. Bendat, J. S.—Piersol, A. G.: Random Data Analysis and Measurement Procedures. New York, Wiley-Interscience 1971.
12. Bergland, G. D.—Dolan, M. T.: Fast Fourier Transform Algorithm. In: Programs for Digital Signal Processing, IEEE Press, 1979, pp. 1.2-1—1.2-18.
13. Blackman, R. B.—Tukey, J. W.: The Measurement of Power Spectra. New York, Dover Publications Inc. 1958.
14. Bluestein, L. I.: A Linear Filtering Approach to the Computation of Discrete Fourier Transform. *Trans. IEEE,* ASSP-**18**, 1978, no. *6*, pp. 451—455.
15. Bogner, R. E.—Constantinides, A. G. (editors): Introduction to Digital Signal Processing. London, John Wiley and Sons 1975.
16. Bracewell, R.: The Fourier Transform and its Applications. New York, Mc Graw-Hill Book Comp. 1965.
17. Brigham, E. O.: The Fast Fourier Transform. Englewood Cliffs, N. J., Prentice Hall 1974.
18. Brodersen, R. W.—Hewes, C. R.—Buss, D. D.: A 500-Stage CCD Transversal filter for Spectral Analysis, *IEEE J. of Solid-State Circuits* SC-**11**, 1976, no. *1*, pp. 75—84.
19. Butzer, P. L.: Fourier Analysis and Approximation. Basel, Birkhauser 1971.
20. Cochran, W. T. et al.: What is the Fast Fourier Transform? *Trans. IEEE,* AU-**15**, 1967, no. 2, pp. 45—55.
21. Cooley, J. W.—Lewis, P. A. W.—Welch, P. D.: Historical Notes on the Fast Fourier Transform. *Trans. IEEE,* AU-**15**, 1967, no. 2, pp. 76—84.

22. Cooley, J. W.—Lewis, P. A. W.—Welch, P. D.: The Fast Fourier Transform and its Applications. *Trans. IEEE,* E-**12**, 1969, no. 1, pp. 27—34.

23. Cooley, J. W.—Lewis, P. A. W.—Welch, P. D.: The Finite Fourier Transform. *Trans. IEEE,* AU-**17**, 1969, no. 2, pp. 77—85.

24. Cooley, J. W.—Lewis, P. A. W.—Welch, P. D.: The Fast Fourier Transform Algorithm: Programming Consideration in the Calculation of Sine, Cosine and Laplace Transforms. *Journ. Sound. Vib.,* vol. **12**, 1970, July, pp. 315—337.

25. Cooley, J. W.—Tukey, J. W.: An algorithm for Machine Calculation of Complex Fourier Series. *Math. of Computation,* vol. **19**, 1965, pp. 297—301.

26. Čížek, V.: Numerische Hilberttransformation. Trans. of the Institute of Radioengineering and Electronics no. 11, Prague 1961.

27. Čížek, V.: Discrete Hilbert Transform. *Trans. IEEE,* AU-**18**, 1970, no. 4, pp. 340—343.

28. Čížek, V.: Accuracy Improvement of Fourier Integral Calculation Using Discrete Fourier Transform. In: Signalverarbeitung — NTG Fachtagung, 4.—6. April 1973, Erlangen—Nürnberg, pp. 286—293.

29. Čížek, V.: Výpočet Hilbertovy transformace užitím diskrétní Fourierovy transformace (Computation of the Hilbert Transform Using the Discrete Fourier Transform). In: Sborník konference o elektronických obvodech (Proceedings of the Conference on Electronic Circuits) 4.—6. 9. 1973, pp. 252—253.

30. Čížek, V.: Použití splinových funkcí k výpočtu Fourierova integrálu rychlou Fourierovou transformací (Application of spline functions to the computation of the Fourier integral by the fast Fourier transform). In: Simulace systémů — Sborník 9. semináře (Simulation of Systems — Proc. of the 9th course) 3.—5. 6. 1975, pp. 1—4.

31. Čížek, V.: Výpočet okamžité frekvence a obálky signálu pomocí diskrétní Fourierovy transformace (Computation of the Instantaneous Frequency and the Envelope of the Signal by Means of the Discrete Fourier Transform). *Acta Polytechnica,* **18**, III, 2, 1977, pp. 33—39.

32. Čížek, V.: Some Generalization of Goertzel Algorithm. In: Proc. of the Summer Symposium on Circuit Theory, Prague 1972, Part II, pp. 56—61.

33. Čížek, V.: Recursive Calculation of Fourier Transform of Discrete Signal. In: Proc. ICASSP 82 — IEEE International Conference on Acoustics, Speech and Signal Processing, Paris 1982, pp. 28—31.

34. Davis, P. J.—Rabinowitz, P.: Methods of Numerical Integration. New York, Academic Press 1975.

35. Ditkin, V. A.—Prudnikova, A. P.: Integral Transforms and Operational Computations. (In Russian) Moscow, Gos. Izd. fiz.-mat. lit. 1961.

36. Eagle, A.: On the Relations Between the Fourier Constants of a Periodic Function and the Coefficients Determined by Harmonic Analysis. *Philos. Mag.* vol. *5*(7), 1928, pp. 115—132.

37. Einarson, B.: Use of Richardson Extrapolation for the Numerical Calculation of Fourier Transform. *Journ. of Comp. Physics,* vol. **21,** 1976, July, pp. 365—370.

38. Erdélyi, A. et al.: Tables of Integral Transform. Vol. 1, vol. 2, New York, Mc Graw-Hill 1954.

39. Fichtengolc, G. M.: A course in differential and integral calculus, vol. III. (In Russian) Moscow, Gos. Izd. fiz.-mat. lit. 1960.

40. Filon, L. N. G.: On the Quadrature Formula for Trigonometric Integrals. *Proc. Roy. Soc. Edinburgh,* vol. *49*, 1928—1929, pp. 38—47.

41. Gautschi, W.: Attenuation Factor in Practical Fourier Analysis. *Num. Math.,* vol. **18**, 1972, pp. 373—400.

42. Gelfand, I. M.—Shilov, G. E.: Generalized Functions. Vol. 1, Academic Press 1964.

43. Gentleman, W. M. – Sande, G.: Fast Fourier Transform for Fun and Profit. In: AFIPS Proc., 1966 Fall Joint Computer Conf., 29, Washington, D. C., Spartan 1966.
44. Goertzel, G.: An Algorithm for the Evaluation of Finite Trigonometric Series. *Amer. Math. Monthly,* vol. **65**, 1958, January, pp. 34 – 35.
45. Gold, B. – Rader, C. M.: Digital Signal Processing. New York, Mc Graw-Hill 1969.
46. Good, J. J.: The Interaction Algorithm and Practical Fourier Series. *J. Royal Statistical Soc., Ser. B,* vol. **20**, 1958, pp. 361 – 372; vol. **20**, 1960, pp. 372 – 375.
47. Gradshteyn, I. S. – Ryzhik, I. M.: Tables of Integrals, Sums, Series and Products. (In Russian) Moscow, Gos. izd. mat.-fiz. lit. 1963.
48. Guillemin, E. A.: Theory of Linear Physical Systems. New York, John Wiley 1963.
49. Hamming, R. W.: Numerical Methods for Scientists and Engineers. New York, Mc Graw-Hill 1962.
50. Hardy, G. H. – Rogosinski, W. W.: Fourier Series. Cambridge, The Syndics of the Cambridge University Press 1962.
51. Hekrdla, J.: Indexové transformace n-rozměrné DFT. (Index transforms of the n-dimensional DFT). Research report Z-1369, Institute of Radioengineering and Electronics, Prague 1983.
52. Hildebrand, F. B.: Introduction to Numerical Analysis. New York, Mc Graw-Hill 1956.
53. Jenkins, G. M. – Watts, D. G.: Spectral Analysis and Its Application. San Francisco, Holden Day 1969.
54. Joyce, D. C.: Survey of Extrapolation Processes in Numerical Analysis. *SIAM Review,* **13**, 1971, October, pp. 435 – 490.
55. Klíma, V.: Souměrné složky (goniometrický interpolační mnohočlen) jako koeficienty Fourierovy řady impulsních funkcí (Symmetric Components [Trigonometric Interpolation Polynomial] as Coefficients of a Fourier Series of Impulse Functions). *Elektrotechnický časopis,* vol. **16**, 1965, no. 9, pp. 540 – 557.
56. Kolba, D. P. – Parks, T. W.: A Prime Factor FFT Algorithm Using High-Speed Convolution. *Trans. IEEE,* ASSP-**25**, 1977, no 4, pp. 281 – 294.
57. Krylov, V. I. – Kruglikova, L. G.: Handbook of Numerical Harmonic Analysis. (In Russian) Minsk, Izd. Nauka i technika, 1968.
58. Krylov, V. I. – Skoblya, N. S.: Methods of Approximate Fourier Transforms and Inversion of Laplace Transforms. (In Russian) Moscow, Izd. Nauka 1974.
59. Kufner, A. – Kadlec, J.: Fourier Series. Prague, Academia 1971.
60. Matyáš, V.: Měření, analýza a vytváření náhodných procesů (Measurement, Analysis and Creation of Random Processes). Prague, SNTL 1976.
61. Mc Clellan, J. H. – Rader, C. M.: Number Theory in Digital Signal Processing. Englewood Cliffs, New Jersey, Prentice Hall 1979.
62. Mc Clellan, J. H. – Nawab, H.: Complex General *N* Winograd Fourier Transform Algorithm (WFTA). In: Programs for Digital Signal Processing. IEEE Press, 1979, pp. 1.7-1 – 1.7-22.
63. Mc Clellan, J. H. – Parks, T. W.: Eigenvalues and Eigenvector Decomposition of the Discrete Fourier Transform. *Trans. IEEE,* ASSP-**26**, 1978, no. *2*, pp. 141 – 150.
64. Morris, L. R.: A Comparative Study of Time Efficient FFT and WFTA Programs for General Purpose Computers. *Trans. IEEE,* ASSP-**26**, 1978, no. *2*, pp. 141 – 150.
65. Nussbaumer, H. J.: Fast Fourier Transform and Convolution Algorithms. Berlin, Springer Verlag 1981.
66. Oppenheim, A. V. – Schafer, R. V.: Digital Signal Processing. Englewood Cliffs, New Jersey, Prentice Hall 1975.
67. Papoulis, A.: The Fourier Integral and its Application. New York, Mc Graw-Hill 1962.

68. Papoulis, A.: Signal Analysis. New York, Mc Graw-Hill 1977.
69. Pease, M. C.: An Adaptation to the Fast Fourier Transform for Parallel Processing. *Journal ACM,* **15**, 1968, pp. 252–264.
70. Pírko, Z.–Veit, J.: Laplaceova transformace (Laplace Transform). Prague–Bratislava, SNTL–ALFA 1970.
71. Rabiner, L. R.–Gold, B.: Theory and Application of Digital Signal Processing. Englewood Cliffs, New Jersey, Prentice Hall 1975.
72. Rabiner, L. R.–Schafer, R. W.: Digital Processing of Speech Signals. Englewood Cliffs, New Jersey, Prentice Hall 1978.
73. Rader, C. M.: Discrete Fourier Transform When the Number of Data Samples is Prime. *Proc. IEEE,* vol. **56**, 1968, no. *6*, pp. 1107–1108.
74. Runge, C.–König, H.: Die Grundlehren der mathematischen Wissenschaften. Vorlesungen über numerisches Rechnen, vol. 11, Berlin. Julius Springer 1934.
75. Salvadori, M. G.–Baron, M. L.: Numerical Methods in Engineering. New York, Prentice Hall 1956.
76. Serebrenikov, M. G.: Harmonic analysis. (In Russian) Moscow, Gos. izdat. tech.-teoret. lit. 1948.
77. Schwartz, L.: Méthodes mathématiques pour les sciences physiques. Paris, Herman 1961.
78. Silverman, H. F.: An Introduction to Programming the Winograd Fourier Transform Algorithm (WFTA). *Trans. IEEE,* ASSP-**25**, 1977, no 3, pp. 239–242.
79. Silverman, H. F.–Pearson, A. E.: On Deconvolution Using the Discrete Fourier Transform. *Trans. IEEE,* AU-**21**, 1973, no 2, pp. 112–128.
80. Singleton, R. C.: An Algorithm for Computing the Mixed Radix Fast Fourier Transform with Arbitrary Factors. *Com. ACM,* vol. **11**, 1968, no. 11, pp. 776–779.
81. Singleton, R. C.: Mixed Radix Fast Fourier Transform. In: Programs for Digital Signal Processing. IEEE Press, 1979, pp. 1.4-1–1.4-18.
82. Sloate, H.: Matrix Representations for Sorting and the Fast Fourier Transform. *IEEE,* CAS-**21**, 1974, no. 1, pp. 109–116.
83. Stoer, J.: Einführung in die numerische Mathematik, Part I. Berlin, Springer Verlag 1976.
84. Tolstov, G. P.: Fourier Series. (In Russian) Moscow, Gos. izd. fiz.-mat. lit. 1960.
85. Veit, J.: Integrální transformace (Integral Transforms). Prague, SNTL 1979.
86. Vernet, J. L.: Storage Capacity and Step Number Reduction by Means of an Odd Discrete Fourier Transform. *Proc. IEEE,* vol. **59**, 1971, no. 10, pp. 1531–1532.
87. Vích, R.: Transformace Z a její použití (Z Transform and its Application). Prague, SNTL 1979.
88. Winograd, S.: On Computing the Discrete Fourier Transform. *Math. of Computation,* vol. **32**, 1978, no. 141, pp. 175–199.
89. Zygmund, A.: Trigonometric Series (Vol. I, vol. II). Cambridge, University Press 1959.

INDEX